Controlling Invertebrate Pests in Agriculture

Jessica Page and Paul Horne

CSIRO

PUBLISHING

National Library of Australia Cataloguing-in-Publication entry

Page, Jessica.

Controlling invertebrate pests in agriculture/by Jessica Page and Paul Horne.

9780643103351 (pbk.)
9780643103368 (epdf)
9780643103375 (epub)

Includes bibliographical references and index.

Invertebrate pests – Control – Australia.
Agricultural pests – Control – Australia.
Pesticides – Application – Australia.

Horne, Paul A. (Paul Anthony), 1956–

632.90994

Published by
CSIRO PUBLISHING
36 Gardiner Road, Clayton VIC 3168
Private Bag 10, Clayton South VIC 3169
Australia
Telephone: [+613] 9545 8555
Local call: 1300 788 000 (Australia only)
Fax: +61 3 9662 7555
Email: csiropublishing@csiro.au
Web site: www.publishing.csiro.au

Front cover: Citrus butterfly

All photographs by the authors unless otherwise stated.

Set in Adobe Minion 11/13.5 and Adobe Helvetica Neue
Edited by Bruce Gillespie
Cover and text design by James Kelly
Typeset by Desktop Concepts P/L, Melbourne
Printed by Ingram Lightning Source

CSIRO PUBLISHING publishes and distributes scientific, technical and health science books and journals from Australia to a worldwide audience and conducts these activities autonomously from the research activities of the Commonwealth Scientific and Industrial Research Organisation (CSIRO). The views expressed in this publication are those of the author(s) and do not necessarily represent those of, and should not be attributed to, the publisher or CSIRO.

Foreword

Insect pests can be a real challenge to farmers as often they can't see them until they are causing significant damage to their crops. This has led to the common response from both farmers and their advisors of using insecticides as the single method to control insect pests. Insecticides are often used with limited knowledge of their true effects on the pest species or their effects on other pest and beneficial insects in their crops. As a result we are seeing insecticide resistance problems, changes in pest importance in crops and loss of a number of the older broad spectrum pesticides from the market. Fortunately we are also seeing new selective insecticides coming on the market and with that, a change of thinking in how to manage insect pests using insecticides, often in association with the use of Integrated Pest Management.

This book introduces the reader to the science and the in-field implementation of Integrated Pest Management and, through a number of case studies across a wide range of crops and environments, will give the reader the confidence that it will work.

Approximately 10 years ago we were grappling with trying to control slugs in arable crops in New Zealand and, through our research, found that none of the chemical slug baits were reliably providing adequate levels of control. A chance meeting with Paul Horne and Jessica Page led us to reconsider the options to manage slugs. This in itself was a challenge for a number of reasons. Firstly traditional methods of research do not lend themselves to evaluating IPM practices, secondly there were few researchers and advisors who had an interest in IPM in broad-acre crops and lastly the challenge of finding farmers who were prepared to give it a go! I think we went through most of the deliberations covered in this book in our quest to understand and deal with just one pest but fortunately we were able to call on Paul Horne and Jessica Page to guide us through the process. As you can see from the New Zealand case study in the book it worked, was rapidly expanded to the whole farm and has now expanded to cover all pests and all crops on a number of farms. Our challenge going forward is to extend the

IPM practices to other farmers and to develop the support capability to help farmers. This book will be an important part of the resources that farmers and advisors will need to help them through the transfer to IPM.

Integrated Pest Management requires a range of behaviours from those involved in IPM. It requires farmers and advisors to consider the pests and beneficial insects from a different perspective and to develop an understanding of the life cycles and ecology of the insects. It needs an understanding of how insecticides work and when and why they don't work, as well as an understanding of the population dynamics of pests and beneficial insects. It requires knowledge of how predators and parasitoids impact on the pest and, very importantly, it requires those involved to observe and monitor the populations of both on their farms. These issues are all addressed within the book. A key requirement is the mind-set of the farmer. The case studies address the farmer's perspective and most importantly the acceptance of risk. The level of risk farmers are prepared to take with regard to insect pests varies with each individual. This book helps de-risk IPM, it clearly shows those farmers who have successfully transitioned to IPM have used skilled professionals to help ensure they make the best decisions, similarly often where they have failed is because they did not have, or seek, adequate knowledge or understanding.

If you are reading this book it indicates you have transitioned beyond the first mind set – that insecticides are the only method to control insects. This book is an excellent resource to help you through the next steps. It cannot provide you with a recipe for IPM but it will certainly provide you with the answers to numerous questions you may have in relation to IPM. Read on and then consider how best to use IPM as a farmer, researcher, advisor or chemical reseller to achieve effective insect control.

Once you have read this excellent resource the next steps are up to you!

Nick Pyke
Chief Executive
Foundation for Arable Research
New Zealand

Contents

Preface and acknowledgements

This idea for this book developed as a result of meeting farmers, agronomists and university students who often had little grasp of the idea that a farm is really an agricultural ecosystem rather than a particular set of crops; and if they did, then how to use that knowledge. We have also found that despite there being good scientists and scientific information available, the information was often poorly communicated; that is, not considered in relation to the many other factors and decisions that farmers need to make.

We also found that, when trying to explain why we were giving particular advice to growers, we were expecting them to have a real understanding of concepts that entomologists and other researchers take for granted. It is not that growers are incapable of understanding the concepts but that these concepts are usually not properly explained. Similarly, the issues around whether or not farmers make change or not are often very poorly understood by researchers. We feel it is important to bridge this gap in knowledge so that farmers and those giving them advice are better able to find sustainable solutions for dealing with invertebrate pests. We think this is essential to be understood by the many groups involved in agricultural research and development if changed practices are to occur. There are enough research projects and case studies that have been undertaken to show the way forward. Anyone denying this information (and there are many) and wanting yet more studies is prolonging a process that is already available and proven. Change can occur now, and is in fact taking place. This book documents this change. We hope that it stimulates further positive change where it is certainly achievable.

We have tried to present, in plain English, a clear view of what is involved in the control of pests, what tools are available and how they can work together or not. We try to explain how pests become problems and how to avoid them instead of trying just to kill them. Because there is a great amount of useable information that is seemingly neglected or unknown, we try to provide some that we have found over the years. No doubt there is more. We hope that we have interpreted a range of scientific information, including issues of making changes to conventional practice, and presented it in a way that can be understood by anyone (including scientists).

Our work with farmers is usually to help them to develop and implement Integrated Pest Management (IPM) strategies, but this book is not about IPM. Although IPM is discussed within the book, the term means a range of things to different people. Arguments between research scientists over what IPM really means can get in the way of helping farmers who want to change practices. If farmers can reduce their reliance on pesticides, that is a good outcome, no matter what it is called. So this book explores the options for pest control available to farmers. In particular we look at whether or not a pesticide is required, and if so, how to decide what to use. We have made the decision to use our knowledge about insects and other invertebrates to help farmers implement change.

First we would like to acknowledge the many farmers and agronomists who have collaborated with us over the years, as without this collaboration we would not have been able to write this book. Our work has been mainly in southern Australia and New Zealand, but the principles involved could be applied to any crop, anywhere. Our company (IPM Technologies P/L) has also applied the same principles to crops in Thailand. The examples of farmers making changes described in Chapter 8 reflect the areas in which we work, and the only exceptions are (a) material on citrus, which is based on information provided to us by James Altmann (Biological Services, Loxton, South Australia) and (b) in avocados and macadamia nuts, where growers and their experiences were referred to us by Richard Llewellyn (BioResources P/L). Dr Philip Keane (Botany Department, La Trobe University, Victoria, Australia) provided all the information on cocoa pod borer presented on cultural controls of this pest.

Some of the examples of implementing IPM were originally commenced through research projects (now completed) funded by industry Research and Development bodies, including Horticulture Australia Limited, AusVeg, Grains Research and Development Corporation, the Foundation for Arable Research (NZ) and the Sustainable Farming Fund (NZ). However, most examples are from our own direct collaboration with farmers and their advisers.

We also would like to thank Dan Papacek, Neil Hives, Lachlan Chilman and James Altmann for discussions on control of pests in different crops over many years, and Janet Horne for help with the manuscript.

Introduction

How do we control invertebrate pests in agriculture? A simple question, and a very important one, given the damage that invertebrate pests can cause. Chemical companies producing pesticides have a very clear answer to this question – along the lines of 'Without pesticides the populations of the earth would starve'. We differ in what we give as our answer. The answer is not easy to give, so it takes a book to give an outline of the different approaches that have been or are currently used and the options available to farmers of different crops around the world. We explain a variety of pest management options available to farmers and the relative merits of each.

We, the authors of this book, spend most of our working lives advising farmers on what to do regarding control of pests. Some farmers listen to us, some listen a bit and some ignore what we say. Some farmers do what we suggest and others do not. In this book we present our thoughts on pest management based on our experiences and look at why some farmers change their practices and why others do not. We present our observations on the options that are available to farmers, and the outcomes of different decisions in both the short and long term. To do this we have divided the book into sections, and have provided examples from our experience in a range of horticultural crops. Our work with pest management has been mainly in southern Australia and New Zealand, so the examples chosen are drawn almost entirely from where we have had firsthand experience. However, we firmly believe that the principles involved could be applied to any crop in any country, so the discussion is applicable in any agricultural situation (horticulture or broad-acre). In tropical areas, for example, there may be more pests but there are also more beneficial species. The book is not a manual for pest management in any particular crop, but the issues discussed can be applied to any crop in any location. Also, our work has been with invertebrate pests and their control, so the term 'pest' in this book means invertebrate pests (such as insects, mites and slugs) and does not include weeds or diseases.

Here we have set out our views on pest management in a range of crops and circumstances, and have tried to describe not only the principles involved but also the practical issues that influence decision-making. There is not one standard 'best' method of controlling a pest in all circumstances, and the choices on what to do are constrained by many different factors, from registration legalities to the weather or time of year. When the full range of options is understood by farmers, there will be an increased range of decisions available for farmers.

This book is written for those who may be interested in developing pest management strategies on their own farms, or for anyone interested in how different strategies can be applied.

To make the reading of this book easier we have used common names of the species being discussed throughout the text of the book, where possible. (Some species do not yet have accepted common names.) For those interested in following this up, in Appendix 1 we have provided the scientific names of all species mentioned, with their common names. In Appendix 2 we list the species by their scientific names.

1

Agricultural ecosystems

If we say that agricultural crops are produced by 'modern methods', usually we mean that many of the same or similar variety of plants will be produced at the one time. 'Sequential plantings' of many vegetable crops means that there will be locally abundant amounts of a range of different aged plants within a region. This also means there will be an abundant to almost limitless supply of food for any pest that feeds on such a crop. More permanent crops, such as glasshouse roses, tree crops and vineyards, present similar situations, offering an abundance of food of the same type for much of the year, if not all year round.

This situation is not new, as farmers have always had to deal with pest problems. What changes are the availability and effectiveness of tools that can be used against pests. Farmers now have an array of synthetic and natural pesticides at their disposal, and can use products, usually insecticide sprays, that have become the basis for most pest management in horticultural crops in Australia and many other countries. Genetically modified crops are also now being used, and in some cases, such as cotton production in Australia, they have become the mainstay for the control of key pests in that industry. Commercially produced beneficial species are also available to counter certain pests.

The use of synthetic insecticides has brought with it its own set of problems, including residues in produce, non-target mortality, secondary pests and insecticide resistance. To counter such problems there have been further developments in pesticide production and in spray management techniques. As will be seen later in the book, some approaches (such as Insecticide Resistance Management strategies) probably have led to on-going problems in pest management while dealing with short-term crises.

Figure 1.1   A boom spray in action

Before modern pesticides and fertilisers

So what did farmers do before the advent of synthetic pesticides and fertilisers, and how do some farmers manage to do without them (or with extremely little use) in highly valuable horticultural crops? The approach to dealing with pests, diseases and soil fertility changed in the late 1940s and early 1950s, when chemical options became widely available. Until then farmers did not have the apparently cheap and easy option of applying chemical products with impressive results. Instead they relied on rotation of crops and the application of composts and mulches: it was necessary for farmers to rotate different crops or pasture carefully on the same paddock. That prevented the build-up of soil-borne diseases, weeds or pests that might affect any one crop type, as the problems could not survive well without a suitable host. Green manure crops and composts have been essential to maintain soil fertility for centuries.

What lives in a crop?

What lives in a crop? Although a monoculture may be grown, it is not the only species living in the paddock. There are *micro-organisms* (bacteria, fungi, nematodes and protozoa) present in all soil: these account for more than all the other life present in terms of biodiversity and sometimes also as biomass. These are beyond the scope of this book, except where some are involved as pests or in pest control.

At a slightly higher level in terms of size are the *micro-invertebrates*: organisms that can only be seen with a microscope. These include some tiny insects and mites, and also insect-like organisms such as springtails. Some of these are pests, some are beneficial in terms of controlling pests, and others are benign, including many species that would contribute to the process of nutrient cycling.

The amount that we know about the number of species of both micro-organisms and micro-invertebrates is very small. Estimates of how many species have been described have recently been estimated using DNA extraction techniques. The results were surprisingly low. Our knowledge of both micro-invertebrates and macro-invertebrates in Australia is far less than in some other parts of the world because of what has been referred to as the 'taxonomic impediment': the relatively large proportion of the Australian invertebrate fauna that is undescribed scientifically. It is extremely difficult for scientists to begin studying individual species or build up the knowledge about particular species because of the lack of information published about them. The situation is different in places such as the UK or Europe, where naturalists and scientists have been observing insects, for example, for centuries, and so a large body of information has been accumulated.

In Australian agricultural ecosystems, of the micro-invertebrates affected by agricultural practices, some will be provided with a more suitable environment while others will be disadvantaged. The impact of pesticides, particularly those applied to the soil, will obviously have great potential to change the species composition of micro-invertebrates in any given patch of soil. However, factors such as tillage, irrigation and rotation can also have massive impacts on the soil fauna.

The *macro-invertebrates* are larger in size and may be defined as being able to be seen without a microscope, or more commonly, can be seen with just the naked eye. Because they are more easily seen we know more about these invertebrates than the smaller species, but still there is much that remains unknown. While more of the larger species have been given scientific names than the tiny ones, that is the limit of our knowledge for many Australian species. Information such as life-cycle duration and ecology, including feeding preferences, are mostly lacking. However, just because we cannot see something does not mean it is not there. Just because organisms do not have a scientific name does not mean that they are not there or cannot function as a biological control agent. Entomologists have long known this, and have conducted experiments called 'predator exclusion trials'. This often involves spraying an insecticide that is broad spectrum and would kill predatory species. Pest species in blocks treated in this way can be compared to untreated blocks and an estimate of the impact of even unknown predatory species can be made.

Predator–prey cycle

In pest management a key concept is the *predator–prey cycle* (see Figures 1.3, 1.4 and 3.1). When numbers of prey (in this case the pest species) are low then numbers of predators that eat them are also low. However, as prey (pest) numbers increase, there is a correspondingly larger food supply for the predators, so the numbers of predators can increase. The larger numbers of predators have an impact on the size of the prey (pest) population and numbers begin to fall. This is the desired result in pest management. The consequence of a smaller supply of prey means that eventually the size of the predator population also begins to drop.

This type of cycle is known to occur with many different types of animals, including lynx and hare, foxes and rabbits. The one given here is western flower thrips and a predatory thrips from strawberry crops in Victoria, Australia. The main difference between these examples is the time that it takes for populations to respond. With mites it is weeks or even days, but for foxes and rabbits it may be months or years that it takes for the populations of each to have an effect on the other.

A second example from our data is control of lettuce aphid in lettuce crops in Victoria, Australia, this time by the predatory brown lacewing (Figure 1.3). Lettuce crops at this time of year in Victoria take about 8 weeks from seedling transplant

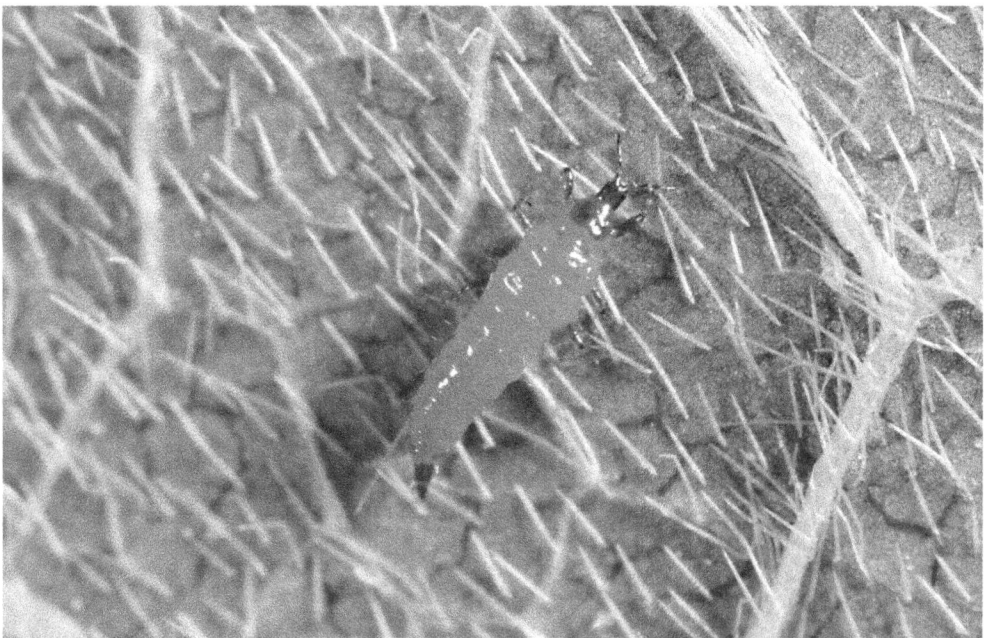

Figure 1.2   Tubular black thrips, a predator of thrips and mites. (Photo by Denis Crawford, Graphic Science)

until harvest. Lettuce aphid is a pest that is tolerant of many insecticides, but also is extremely difficult to target within head lettuce (such as 'iceberg'), because it is protected from direct contact with insecticides because of the structure of the plant. The aphids live inside the heart of lettuces and are protected by overlapping leaves. The accepted method of dealing with the pest in Australia and New Zealand has been to drench the seedling plug in the nursery, just before planting in the field, with a high enough rate of the systemic insecticide Confidor to protect the plant from lettuce aphid for the life of the crop (until harvested).

This method requires a decision to be made, before the crop is planted into the paddock, as to whether or not the insecticide treatment is to be used. Some varieties of lettuce are resistant to lettuce aphid, but these are not always considered by growers to be as suitable as susceptible varieties. Biological control agents (such as brown lacewings, hoverflies and ladybird beetles) as part of an IPM strategy can deal very effectively with lettuce aphid. Commercial crops are grown successfully every year in Australia without insecticide drenches.

One problem is the short time that it takes to grow the crop; another is that the decision on control needs to be made before planting. If the decision is made to *not* use insecticide drenches on seedlings, it is made because (at the time) there were no insecticide sprays that would be effective. (This situation has since changed with the arrival on the market of Movento, which is a true systemic insecticide.) The short life of the crop is sometimes seen as being too short for biological control to work, because the predators do not arrive before their food source (lettuce aphid).

Figure 1.4 illustrates what we have seen happen in lettuce crops, but particularly the first planting when moving to a new paddock where no predators of lettuce aphid are already present. It shows how lettuce aphid arrives very soon after lettuce seedlings are planted, but there is a considerable lag before the predators of lettuce aphid (in this case brown lacewings) begin to increase and have an impact on the lettuce aphid population. However, once the brown lacewings begin to breed within the lettuce then the subsequent control of lettuce aphid, as measured by the extremely rapid drop in lettuce aphid level, is complete and achieved before the crop is harvested. Lettuce aphid is a problem as a contaminant rather than causing direct damage and so once the aphids are eaten there is no residual damage and as there is no food for the lacewings (or other predators) then they also leave in search of a new food source.

These two examples illustrate how effective biological control agents can be, but the worry for growers is the level of pests that is reached before control occurs, and trusting that control is in fact going to occur. For effective control, the height of the peak in numbers of pests (prey) should be tolerable until the numbers of predators build up to a level that will achieve control. The shape of the graph in both cases is the same: a peak of prey followed by a peak of predators, and then a drop in both.

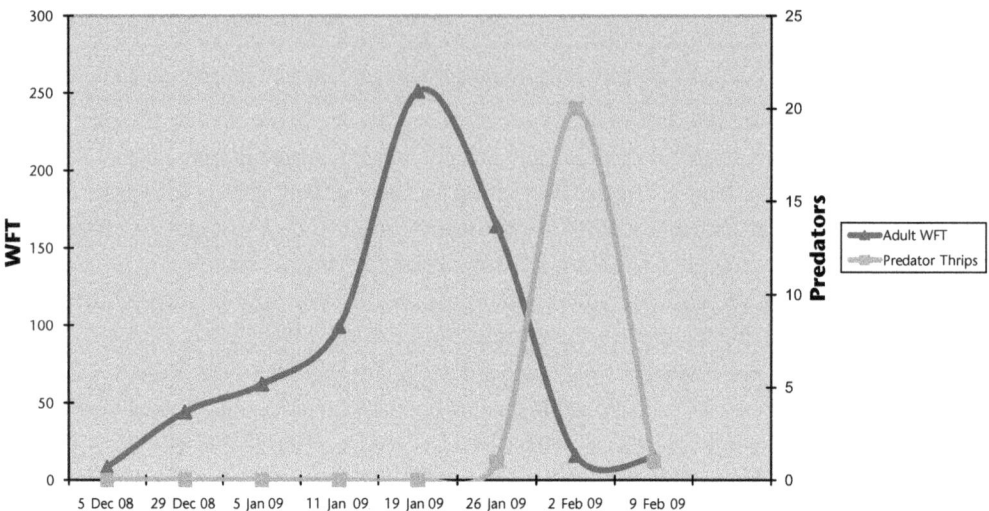

Figure 1.3 Numbers of western flower thrips and predatory tubular black thrips (in strawberry crops, Victoria, Australia)

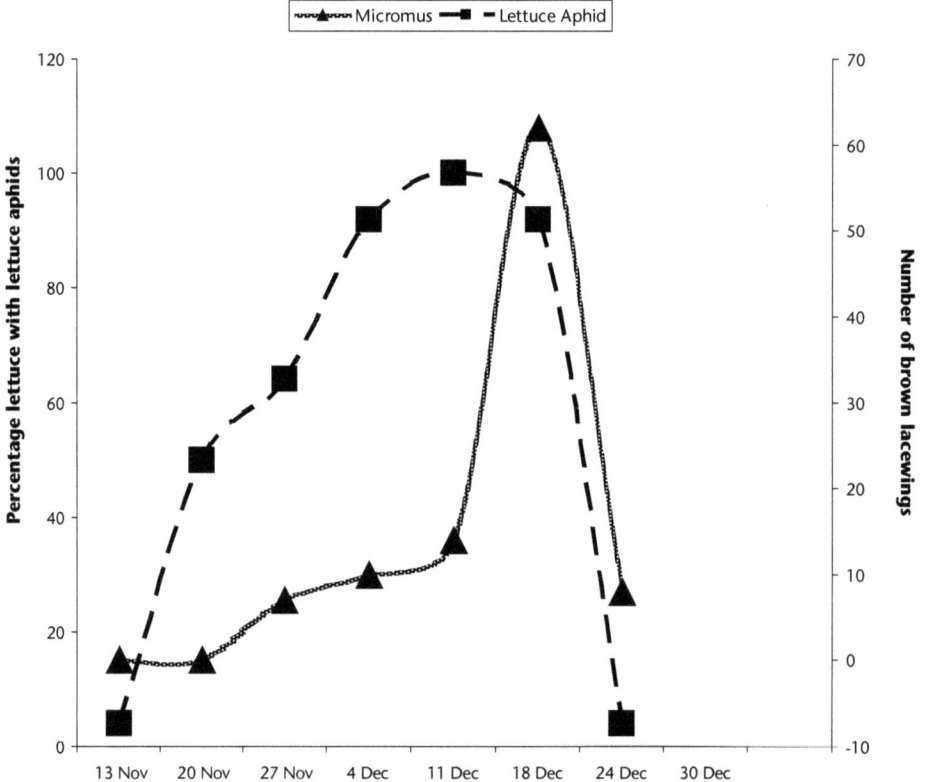

Figure 1.4 Lettuce aphid control by brown lacewings (*Micromus*)

Another example of this type of interaction that we often encounter is control of two-spotted mite by another mite called *Persimilis*. Two-spotted mite is a serious pest for many horticultural crops, including fruit (such as apples, pears and strawberries), flowers (roses and gerberas, including glasshouse and outdoor crops), nursery plants (ornamental horticulture) and some vegetables (such as beans, capsicums, cucumbers and eggplants), but it is often a major pest in glasshouse crops. This pest can cause serious damage by feeding on the cells of host plants, and if in high enough numbers, populations of the mite can defoliate plants. The foliage appears to have been burnt, because the cell contents are destroyed. When the pest populations are high they cover leaves, or what is left of them, with webbing, which gives rise to a commonly used name of spider-mite.

A problem for farmers trying to deal with two-spotted mite is that, in many cases, the pest is resistant to most of the miticides available. Although the newer miticides may work, if overused they fail because the short generation time of two-spotted mite (particularly in warm, dry conditions) allows resistance to develop very rapidly. However, as mentioned above, two-spotted mite has a specific predator, the mite called *Persimilis*, and it cannot rapidly develop resistance to this natural enemy.

Persimilis is more susceptible to pesticides (particularly insecticides and insecticide residues) than two-spotted mite. For example, a single application of some insecticides can kill *Persimilis* if it walks on treated leaves for 12 weeks or more after the original application in glasshouse crops. However, if there are no complicating factors as a result of pesticides, then *Persimilis* can be introduced, as it is commercially available. If two-spotted mite populations are large, they provide a large food source for *Persimilis*, so the predatory mites begin laying many eggs among the pests. At first the predators may be greatly outnumbered and the pest population will continue to grow. However, as each generation produces ever-increasing numbers of predators, they begin to have an impact on the pest population, which at first stops increasing in size, then begins decreasing. This lag between the initial introduction of the predator and the point of achieving control depends mainly on the starting size of the pest population and the number of *Persimilis* introduced. Factors such as temperature and humidity will influence the rate of population increase or decrease, but the relative number of pests and predators is the most important factor. When the population of two-spotted mite is controlled, then the *Persimilis* do not have a food source, so their numbers drop. After balance is achieved, both species are still present, but the level of damage stays below that of economic loss.

For many farmers who have had to deal with two-spotted mite by the use of miticides, achieving control with *Persimilis* can be an astounding achievement: one that they want to avoid disrupting by using redundant pesticide applications *aimed at other pests or diseases.*

Understanding the type of interaction

Why is it important that this type of interaction is understood? Mainly because we need to consider pests as populations, not just as individuals, as in most cases pests do not arrive (or leave) overnight, although at times it may appear that way. (Obvious exceptions are pests such as locusts, Rutherglen bugs and plague thrips, which actually do arrive overnight.) A pesticide-based approach may seem successful, as it kills large numbers of pests causing immediate damage, but it needs to be remembered that these individuals are only a part of a population that interacts with populations of other species (beneficial and pests). So although short-term problems may be solved, a sustainable strategy would also take into account populations of beneficial species and other pests. It is best to view a crop of any type as an agricultural ecosystem, not as a sterile laboratory. This is something that wholesale buyers (and individual consumers) also need to remember when they put the onus on the farmer to deliver to them a product that is free of any insect (dead or alive).

So the issues discussed in this book are not only how to deal with individual pest species in the current crop, but how to achieve good management of pest populations in the current crop and in a sustainable way in the future.

2

Pesticides

The use of synthetic pesticides to protect crops from pests, which is now a standard tool available to farmers, has for the past 50 to 60 years been the mainstay of protection for most crops. What has changed more recently is the frequency, manner and type of pesticides used, and the relative importance of other methods used alongside pesticides, particularly for control of invertebrate pests. For example, instead of routine, scheduled applications of broad-spectrum insecticides, farmers may now choose to use selective insecticides only when required, as determined by monitoring. They may use seed dressings instead of boom-spray applications to minimise amounts of pesticide applied and non-target mortality. They may rely more on biological control agents (natural enemies of the pests) than on pesticides. What we have seen is an increasing set of options for farmers in terms of pest management and an increasing willingness to use more than a cheap insecticide spray.

The availability of synthetic pesticides in the 1950s changed the way most farmers dealt with pests. A simple application of a pesticide had dramatic impact on target pest populations, and the products developed dealt with a range of pests. New insecticides, miticides, fungicides and herbicides were developed to deal with particular types of pests, but in general they dealt with many species; that is, either they killed sucking insects or chewing insects or killed on contact. The use of these products became accepted as the standard way to deal with pests and the main decision farmers had to make was to choose which product to spray.

There has been, and still is, a large amount of time, money and effort spent on spray technology. There have been tremendous advances in this area, and the technology has enabled growers to change droplet size to suit conditions and reduce drift, apply precise amounts on the crop in the field or nursery with computer controlled application methods, and select nozzles to suit particular crops, speed of application or target pests. These advances have been made because spraying pesticides is the main method of protecting crops, so there is a ready market for good products. Our concern is with what is applied more so than how it is applied, but when a pesticide is used, whether within or outside of an IPM strategy, it needs to work well.

All pesticides used in Australia are now required to be registered by the Australian Pesticides and Veterinary Medicines Authority (APVMA). Any product being sold with claims that it kills pests is considered a pesticide, and requires approval from the APVMA before it is allowed to be used in Australia. There are some differences in state laws regarding the use of pesticides on specific crops, and most states require a permit for a product to be used on a particular crop if the company producing it has not sought registration for use on that crop. For example, a chemical company may seek to have a product registered on a major crop such as potatoes (where the potential market is large), but may not want to pay for the significant cost of registering the same product on a whole range of smaller crops (such as herbs or capsicums). In the latter case, the states may require permits for the use of the pesticide on minor crops.

So the use of pesticide is regulated, and rules are set out on the minimum length of time required before harvest following the application of each pesticide (the *withholding period*). This time may be as little as a day, or may be weeks or months. *Re-entry periods* (the time that must elapse following a spray before workers or anyone else can re-enter the crop) are sometimes regulated, but in practice there is very little information or enforcement of re-entry periods. There are also guidelines provided by the chemical companies or other groups about the maximum number of times a product should be used on any particular crop. Chemical companies also have 'stewardship' guidelines that attempt to influence the use patterns of their products (particularly the new products on the market). Despite this regulation, there are some recurring problems that farmers face with the use of pesticides. These are very real problems that farmers have to deal with regularly, and they have to weigh up many different factors when considering pesticide use. These are discussed below.

Insecticide resistance

Insects and mites can develop resistance to pesticides by different methods, but basically the individuals within a population that are able to cope with a certain dose of pesticide (when other more susceptible individuals are killed) will be the only ones to pass their genes on to the next generation. That is, each application of the pesticide selects for those individuals that can survive that treatment. In this way the genetics of the population shifts, so that instead of a very small number of individuals being able to tolerate a pesticide, most individuals in the population become resistant to that dose of pesticide. The mechanism of resistance can vary from being able to detoxify the poison within the insects' body to simply developing more protection (such as thicker cuticle) or changed behaviour. So, although initial results can be very dramatic, with very high rates of kill, the development of resistance begins without any obvious signs, as the percentage of individuals that may survive is usually very low in the initial stages. Some insects and mites, especially those with short life-cycles, can develop resistance very rapidly. For example, western flower thrips and diamondback moth have generation times of less than 21 days in high temperatures. When warm to hot temperatures are normal and pesticide applications frequent, then insecticide resistance in these and similar species is assured.

When the selection pressure is removed (that is, the pesticide involved is no longer used) it is possible for the population to gradually shift to having a higher percentage of susceptible individuals. When this happens and the same pesticide is used again, the results again appear to be good, but the population is immediately transformed into being composed of resistant individuals, and the pesticide will not give the same desired result a second time if used soon after the first.

Cross-resistance

Pesticides are categorised into different groups based on their chemical structure. Similar pesticides may be developed by one or more chemical companies but they can have very similar chemical structures. Some of the major chemical groups that formerly dominated the insecticide and miticides markets over the years are the organochlorines, organophoshates, carbamates and synthetic pyrethroids. While products from these groups are still sold, there are now many other groups, including the neo-nicatinoids and fiproles that have been developed (see Table 2.1).

When insects and mites develop resistance to a particular pesticide, then they have developed the means to deal with a particular chemical. If they have developed a way to detoxify or bind a particular molecule, then they are more easily able to develop resistance to closely related products, because the molecules are very similar. This is referred to as *cross-resistance*. Management of pesticide resistance is discussed later in this chapter.

The problem of insecticide resistance was noticed very soon after the development of the first of the widely used synthetic insecticides, DDT. Entomologists found that house flies had developed resistance to DDT by 1950. It is a real problem, both for farmers and the companies producing the pesticides, as the products farmers had been used to using to protect their crops no longer work, so they stop buying them. The choices available to a farmer who is used to using a

Table 2.1 Different chemical groups and effects on beneficial species*

Chemical group	General effect on beneficials	Beneficial species adversely affected
Organochlorines	Bad	All
Organophosphates	Bad	All
Carbamates	Variable but mostly bad	Wasps and flies
Synthetic pyrethroids	Bad	All
Neo-Nicotinoids	Bad	Many beneficial species
Fiproles	Bad	All
Spinosin	Variable	Wasps

* There is no general evaluation of the variable pesticides on beneficial species. There is a need for information on the effect of each pesticide on each beneficial species, and indeed each life-stage.

pesticide-based approach and now must deal with pesticide-resistant pests are usually to increase the rate or frequency of application of the product, change to a different product, or try something totally different in terms of pest management.

Residues in produce; withholding periods

To protect consumers from buying and consuming food with excessive chemical residues, there are laws that regulate the minimum time after any product has been sprayed that the crop can be harvested. This information is contained on the label on each container of pesticide. The measure that is used to assess whether or not there is more than acceptable levels of pesticide is to determine if chemical residues are within *Maximum Residue Levels (MRL)*. If a chemical is permitted to be used on a crop, then there will be a maximum allowable residue (usually extremely small, such as 0.01 ppm). However, if the product is not registered or permitted on a crop, then the MRL is zero. The accuracy of chemical testing is extremely high, and is able to detect such tiny amounts of particular chemicals. However, how the MRL relates to food safety is up for discussion. For example, the MRL for the same product in citrus and lettuce is as follows: citrus has an MRL of 0.2 mg/kg, while lettuce is 5.0 mg/kg (a leafy vegetable that is consumed entirely). Obviously there is room for discussion here!

Government agencies conduct various types of testing for chemical residues in produce, including a *market basket survey*, where food items are randomly selected from where they are sold to consumers and tested for a range of pesticides. However, it is not only the government agencies that conduct such tests. Supermarkets also conduct tests on products that they buy from farmers, as do some independent groups, such as Greenpeace and the Australian Consumers Association, the organisation that publishes *Choice* magazine. Food being exported from Australia may also be subject to routine testing for chemicals, some of which are permitted to be used by Australian law but not by overseas countries (such as the European Union) on the same crop.

So Australian farmers are required to be very careful with their use of pesticides to meet these requirements, and testing results show that overall the chemical residues are within the allowable limits. The limitations mean that the selection of pesticides is not based just on what will work against the pest, but on the timing of harvest and other standards that will be applied. In short, when growing crops where the time from planting to harvest is brief, there can be greater reliance on pesticides with a short withholding period. This has implications for resistance management and problems with secondary pests (see below).

The detection of pesticide residues in food has had massive implications for pesticide use and regulation as well as availability of pesticides in Australian agriculture. For example, in 1987 residues of organochlorine insecticides (in particular, dieldrin and DDT) were detected in meat that was exported from Australia to the USA and Japan. The immediate effect (following the rejection of the shipments and a ban on Australian meat) was that organochlorine pesticides were immediately banned from all agricultural uses in all Australian states; that is, the lengthy list of permits for particular uses of all organochlorines was immediately cancelled. This included the related products of heptachlor and lindane, but the availability of the non-persistent product endosulfan has continued until now. We expect it to be unavailable in Australian agriculture in the near future.

A similar event happened in the 1990s when a product called Alar was found in export beef from Australia. The product had been used in horticulture; the crop trash had been fed to cattle in a drought year when normal fodder was in short supply, so the meat was thus contaminated. As a result, the company that produced the pesticide decided to immediately stop producing and selling the product in Australia. Therefore the product was not available to those horticultural users who had found it to be a useful product within IPM strategies. There are examples of chemical companies (such as Shell and ICI) deciding to stop producing pesticides altogether; these production bans are often related to the problems associated with perceived or real problems with pesticides other than their effectiveness on the target pests.

Non-target mortality

The concept of non-target mortality is really simple, but the ramifications of it have been greatly underestimated and overlooked.

Non-target mortality simply means that a pesticide applied to kill one pest has killed one or more other species in addition to the target. We do not mean that the pesticide has been washed into rivers causing fish kill, or that drift of pesticide has meant that the spray has ended up somewhere other than it was intended. While these situations have occurred, instead we are discussing situations in which the pesticide has been applied properly to the target crop. A pesticide may kill more than one pest but it may also kill other species that are considered beneficial or benign. For example, DDT applied to a crop to kill (say) caterpillars would kill other pests, such as aphids, but also kill parasitic wasps that are biological control agents of aphids and caterpillars. The bio-accumulation of DDT through the food chain meant that predators higher up the food chain (such as bald eagles) were being affected more than species lower in the food chain.

Rachel Carson, in her book *Silent Spring* (1962), was the first to alert a wider audience to the impact of synthetic insecticides on non-target organisms. The examples that Rachel Carson used were of the poisonous effects on wildlife because of bio-accumulation of pesticides through the food chain, seen most intensively in the bodies of the top predators. Thinning of eggshells was one perceived impact of pesticides on non-target species. Obviously the manufacturers of DDT did not intend to thin the eggshells of bald eagles, but that was the outcome. The impacts of pesticide applications are not immediately obvious and cannot always be predicted.

Regulators (such as APVMA in Australia) require pesticide companies to provide information on some non-target species, but these relate to only a very few species that are of general interest, such as earthworms and bees. It is impossible to insist that chemical companies provide all information on all possible effects on all species. If we recognise this, we should also accept that registered pesticides are not automatically certifiably 'safe'.

As a result of our own testing, IPM Technologies Pty Ltd has concluded that no synthetic chemical product is safe to all species. Although some products are safe to most species tested, they still have detrimental effects on particular species. The list of detrimental effects is likely to increase as testing increases.

Despite the statement above, there is a great difference between different pesticides and pesticide groups. The early synthetic insecticide groups such as the organochlorines, organophosphates and carbamates, followed by the synthetic pyrethroids, were generally expected to kill a wide range of species. Classified as 'broad-spectrum', these pesticides were designed to kill a wide range of pests, but they also killed a wide range of other (beneficial) species. There are exceptions to this general statement; for example, pirimicarb (Pirimor) is a carbamate insecticide that is relatively selective when applied to control aphids. It does kill other species, especially parasitoid wasp species, but is highly volatile, so the effects on non-target species occur only for a very short time after application. This is in contrast to other pesticides in the same chemical group (such as methomyl or carbaryl), which have highly broad-spectrum effects, and for a long time after application (especially in glasshouse situations). (See Chapter 2, 'Selective insecticides'.)

Disruption of biological control (primary and secondary pests)

It may be hard for some to accept, but a single application of a pesticide such as methomyl in a glasshouse crop such as roses would effectively prevent the establishment of key beneficial species such as *Persimilis* (a predatory mite that eats two-spotted mite) or *Encarsia* (a wasp that attacks greenhouse whitefly) for 3 to 4 months after that application. During that period there will be no prospect of biological control of key pests. *Persimilis* preys on two-spotted mite, and *Encarsia* parasitises greenhouse whitefly; it is essential that these species can survive and prosper if biological control is to occur. Given that a single spray of an insecticide such as methomyl would prevent biological control from working means that, if applied, further insecticides would be required because the previous biological control had not worked. In this way, continued insecticide applications become viewed as essential.

The dilemma for the farmer in this situation is that he or she knows that the pesticide applied will prevent biological control, but if it is not applied there will probably be an immediate increase in pest numbers.

Several different and complex relations can be examined. The first is the most obvious, where primary biological control agents are killed by the application of an insecticide. That is, the most important biological control agent of a particular pest is disrupted (killed) by a pesticide applied for other pests or diseases, for example, the control of two-spotted mite, and its natural enemy *Persimilis*. Two-spotted mite is very common in horticultural crops, including glasshouse crops, but also in crops such as apples, pears, grapes and strawberries. It is resistant to many insecticides and miticides, so control of this pest is critical for many farmers.

In this first example there is a primary pest and a primary biological control agent. If a pesticide is applied, killing or otherwise disrupting a key biological control agent, then the pest will have an advantage. If the pest can survive the pesticide that is applied, that species will have been given a great advantage, as it will have no natural enemies left alive, no pesticide that will worry it and an unlimited food source. In this way a pest can develop from being insignificant to extremely serious.

A second method of pesticide disruption to pest management is less obvious, but is no less significant, especially in broad-acre cropping at present. Pesticides are applied to control a pest, and this can be very successful, but the application induces secondary pests that are not so easy to control. One example of this is diamondback moth, which is a serious pest of brassica crops around the world. It has been able to become resistant to insecticides, and was induced by sprays that were originally targeting cabbage white butterfly. The cabbage white may be controlled, but it has now been replaced in importance by diamondback moth, simply because of the differences in tolerance to insecticides.

Another example is pesticides applied to control redlegged earth mite (RLEM), but causing problems with other pests, such as lucerne flea, and other mite species, such as blue oat mite. In this case there has been the development of a scientifically sound method of dealing with RLEM by CSIRO. It developed a method called the TimeRite strategy.

TimeRite

The target pest in this case is a mite: redlegged earth mite. This is a pest of broadleaf plants, in particular in broad-acre crops and pasture, so canola and clover are particularly susceptible. Normally this mite would produce an over-summering egg that would survive the hot and dry conditions, then emerge in the autumn after rains. Researchers showed that this species does not survive over summer except as an over-summering egg, and therefore, if the adult females that produced this resistant stage could be killed, the damaging autumn population of pest mites would not be present.

This strategy has been very effective, except that the pesticides used kill beneficial species of insects and mites as well as pest species. Because of this strategy, beneficial species that would otherwise control not only redlegged earth mite but other pests are killed, so farmers become locked in to spraying insecticides to control an ever-increasing number of pests. For example, sprays targeting redlegged earth mite also kill bdellid mites that prey on lucerne flea, so lucerne flea becomes a worse problem. Spraying for pests such as lucerne flea, blue oat mite and other mites is becoming more common, with these latter pests becoming even harder to kill, so higher rates are required. Obviously this sequence of events is not desirable, and it would be better to avoid inducing such pests. The savings involved in never having to bother with such pests are immense but often not considered. Too often only one pest is considered, and the impact of any spray on the loss of control for other pests is overlooked.

Insecticide resistance and beneficial species

Pests becoming resistant to insecticides is a common problem, as described above. Why do beneficial insects not become resistant as well, and, more importantly, what implications would this effect have for pest management? There have been attempts over the years to induce insecticide resistance in beneficial species, usually without success, but there have been some examples where this was achieved.

Predatory insects and mites eat more than one prey individual, so can pick up higher levels of insecticide than their prey as they accumulate a dose from each treated prey. This is called *secondary poisoning*. For example, a carabid beetle that is a scavenger as well as a predator could pick up a lethal dose of pesticide, even when it was not contacted directly, by feeding on caterpillars that have been killed by insecticide spray, or by eating dead slugs that had been poisoned by carbamate insecticide-based pellets.

Usually the dose of a broad-spectrum insecticide that is used to kill pests will also be more than sufficient to kill most beneficial species that are exposed to the pesticides. However, some examples (such as control of scale insects in citrus) show how hard-to-kill pests with pesticides can be made even harder to control when their natural enemies (wasps) are killed by the same pesticides.

However, many of the newer insecticides are much less predictable in their effects, which is discussed in the next section. One example of a beneficial mite being tolerant of insecticides is where *Persimilis*, a predator of two-spotted mite, was commercially reared and made tolerant to the insecticide carbaryl. This allowed carbaryl to be applied in crops without losing control of two-spotted mite, because *Persimilis* was able to survive. The problem with this approach is that, although it may seem like a good thing, the carbaryl applications will still kill a whole range of other beneficial species, so control of other pests will be lost. It is an example of where the focus has been on one pest, not integrating control measures for all pests.

Just as there are different tolerances to pesticides by pests, there are also different tolerances by beneficial species. In the example above, *Persimilis* is the main biological control agent for two-spotted mite in most cases. However, it is more affected by pesticides and also high temperatures than another predatory mite, which we call Mite B (*Neoseulius wearnii*). So in conditions where there is some disruption of *Persimilis* by either pesticides or high temperatures or both, Mite B is given an advantage over *Persimilis*, although it is usually the other way around. The presence of Mite B as the dominant predator of two-spotted mite in a crop can be used as an indicator that something is disrupting *Persimilis*.

Selective insecticides

Much of the discussion about insecticides has been concerned with broad-spectrum insecticides that were produced to kill as wide a range of (pest) species as possible. These have caused problems, as described above. However, there are also pesticides that are not broad spectrum and do not kill either all pests or all beneficial species. Some pesticides are highly selective, and kill only a very few target species. For example, insecticides that are actually viruses have been produced to control key pests in some crops; notably, the products GemStar and Vivus are insecticides based on a virus that only kills caterpillars in the genus *Helicoverpa* (often more commonly known as *Heliothis*). Also, products based on a toxin produced by the bacterium *Bacillus thuringiensis* sub-species *kurstacki* and *aizawai* are toxic only to caterpillars. These are sold under many different trade names, including Dipel, Delfin (which are *B.t. kurstacki*) and XenTari (which is *B.t. aizawai*), and are commonly referred to as BTs.

The virus- and bacterial-based products are the most selective insecticides available, but lead to practical problems if farmers want to use them. The main one is that they are vulnerable to breakdown by ultraviolet (UV) light. As Australia has plenty of UV light in the warmer months, there are a range of additional measures that farmers must take to minimise the breakdown and ensure that the sprays are effective.

Surprisingly, these products are highly selective, so they will not kill other pests. GemStar, for example, will kill *Heliothis*, but not other caterpillars that look like *Heliothis*. So if there are a couple of species of caterpillars present at the same time, farmers (or their advisers) need to be aware of this fact, then either use a product that will deal with both or use both virus and bacterial products.

As referred to in earlier sections of this book, newer insecticides are far less harmful to (many) beneficial species than the older broad-spectrum insecticides. However, none that we know of (and we have tested many) is harmless to all beneficials. We need to describe here the different ways of testing pesticides. Chemical companies aiming to produce a pesticide that kills particular pests need to prove to Australian Government regulators that it is actually effective in controlling the nominated pests, and also that it is safe to humans and a selected group of invertebrates, such as honey bees and earthworms. However, there is currently no requirement for testing to be done on the beneficial species, discussed in this book, that are biological control agents for particular pests. The companies producing new chemicals often commission research to determine the effects of their products on beneficial species, but the information produced is owned by the companies, and the release of information is obviously up to each company's discretion.

Independent research on the effects of pesticides on beneficial species is also conducted by a range of organisations, including ourselves. Usually such research is commissioned by industry-based organisations, such as Horticulture Australia Limited (HAL) or AusVeg, to provide information to groups that have identified that such information is necessary to their decision-making on pest control. Information on the effects of pesticides on some commercially produced species of beneficial insects and mites is readily available from websites of companies such as Koppert and Biobest, but the data are applicable only to the species tested. The species tested are often not species that occur in Australia.

Side-effects testing needs to be species specific, and it needs to include more than acute toxic effects; that is, the testing on beneficials needs to be different in design and assessment than that done on pest species. What is needed is an assessment of whether or not a product is safe to particular beneficial species that are important in particular crops. To describe a pesticide as 'safe' to us means that it is neither acutely toxic, nor does it have sublethal effects.

Sublethal effects of pesticides

Sublethal effects are those that are not immediately obvious. Invertebrates can survive the spray (or other application) and so an acute assessment of mortality would indicate that it is 'safe'. However, if the pesticide has effects on fecundity (how many eggs or offspring are produced), on lifespan or behaviour, for example, this would not be immediately apparent but could be just as great. Examples include fungicides such as mancozeb, which reduce egg production in some predatory mites by around 70%. Similarly, work in our laboratory has found that the insecticide pymetrozine is safe to almost all predators tested, but if sprayed on first instar ladybird larvae, they will develop until the pupal stage, and then not emerge as adult beetles. In both of these cases it is the *population* of predators that is affected. Observations shortly after the application of the pesticides would not pick up these effects, but the information needed before a farmer can decide on which pesticide is safe or not will require the longer-term test results.

When using such results it is important to remember that they are usually based on one single spray at the label rate. If the pesticide is used repeatedly or at a higher rate than was tested, then the results are likely to be different (not safe). Information is currently available on-line from organisations such as Koppert, Biobest and AusVeg. However, not all testing is done to the same standard (and often does not include sublethal testing), so the results need to be used cautiously.

When products are identified as being safe to the particular beneficials in a particular crop, then the problem of overuse can lead to resistance. That is, if one pesticide is identified as safe and is the only one that is regarded as such, obviously the farmer using an IPM approach would not want to kill key beneficial species, and so may have overreliance on the one compatible IPM product. If it is used too often, resistance will develop, so the best products for IPM will be lost. We emphasise that simply using selective insecticides as the basis for control is not IPM.

'Safe' insecticides?

As may be inferred from the above discussion it is extremely difficult to say an insecticide (or any pesticide) is safe or not, as it depends on many variables, including the crop, the species of beneficial and the timing of application. However, one important aspect that has not been discussed is the relative numbers of pests and beneficials at the time of application. This can be extremely important in determining whether a pesticide application is safe or unsafe.

It is important to take into account the risk of *pest flare*. This situation occurs when something (usually a pesticide application) kills beneficial species that would otherwise have kept the pest under control. An example of this is shown in Chapter 8 (potatoes), where the beneficial species controlling aphids were killed and the aphid population flared up. In a situation where key pests such as two-spotted mite and western flower thrips are under control, then a single application of a pesticide that will kill the predators of these pests is unlikely to cause immediate problems. The very low numbers of the pest species means that they are not able to increase to damaging levels within one generation, so there is the possibility of restoring the balance, either by releasing beneficial species or relying on naturally occurring immigration of beneficials. However, in a different scenario where pest levels of these two species are high, then an application of the same pesticide will mean that there is an immediate damaging level of pests. Therefore there will be damaging levels of pests within one generation and an increase in pest numbers to a point where biological control will be slow and so continuing damage could occur.

So the same pesticide applied to the same crop could be safe in one scenario (when key pest levels are low) or could be highly disruptive (when key pest levels are high). Once again, we emphasise that pesticide applications need to be considered on a site-specific basis, not as a general rule. What is safe in one situation could be disastrous in another.

Insecticide Resistance Management Strategies (IRM)

One technique to slow down the development of resistance is the use of Insecticide Resistance Management Strategies (IRM): only one letter different from IPM but a fundamentally different approach. It relies on a farmer not using only one chemical or chemicals from within the same chemical group (see section above on 'Cross-resistance'), but instead rotating through a range of products from different chemical groups. In this way the pests are not constantly exposed to the same chemicals, so populations will not build up resistance so rapidly.

One example of such a strategy is that promoted by some researchers in Australia who have worked on control of diamondback moth. It divides the cropping season into two different time zones, and suggests that growers select only from the chemicals nominated for Time A for that period, then select only from the chemicals nominated for Time B for the second period. In this way the insect pest is exposed to the same chemical groups only a relatively small number of times, so the development of resistance to each component chemical in the strategy could be expected to be slowed.

The problem with this strategy is that there is no discrimination between pesticides that are safe to beneficials or are highly disruptive to beneficial species. There are extremely few chemicals that can be used safely within any IPM strategy, as most pesticides kill at least some beneficial species (see section above on 'Safe insecticides'). Even if a few selective pesticides are used, if one kills wasps, another kills damsel bugs and another kills ladybirds, then rotating through these products will leave the populations of beneficials severely reduced. Therefore, if an IRM strategy is adopted, an IPM strategy is almost certainly forfeited. That is, there needs to be a decision by the farmers as to whether an IPM or IRM strategy is to be adopted.

This is obviously not something that many farmers would have thought about. What each farmer wants to know is what is safe to spray and what is the best method of controlling pests. Learning the difference between IRM and IPM is something that is not high on most farmers' priority lists. What is needed is a simple method of explaining the positive and negative effects of each choice.

Zero tolerance

We often encounter the proposition that 'We have zero tolerance' to pest presence or pest damage. This is often imposed by the buyer. That may be so, but what is not reasonable is the very common assumption that the only way to achieve this is by the routine application of insecticides. Quite the opposite is often true. In many cases, pest problems are made worse and new and more difficult-to-control pests are induced. If the argument of zero tolerance is proposed, it should be linked to an examination of the production system involved and ways of dealing with the range of pests that is present.

A real example of how 'zero tolerance' affects grower practice is when buyers (for example, buyers for supermarkets) reject deliveries of produce when there is any insect found in a shipment. This may be because a ladybird beetle or a fly was found, even though these are not pests that cause damage, or even pose a risk to the consumer. It is easier for the buyers to say 'No insects' and leave it up to the farmer how to achieve that. The crop may have been grown without any insecticides being applied (favoured by most consumers, we expect), but the farmer is forced to apply insecticides immediately before harvest to satisfy the buyer's demands. So the tiny minority of consumers who may complain about a ladybird in their vegetables are causing insecticide use that the vast majority of the population would not want. An outcome of this approach sometimes is that produce is rejected because it contains dead insects. So the farmer is in a difficult position in trying to meet artificial standards.

A second example is where, in attempting to achieve 'zero tolerance' standards, the number of pests is actually made to increase. Real examples of this occur where sprays of insecticides targeting a pest cause insecticide-resistant pests to flourish. Spraying for thrips in leeks induced two-spotted mite problems for a leek grower, who was exporting the produce to a zero-tolerant destination. In fact, there were fewer pests in the crop when he stopped spraying insecticides on the leeks. (For the full story, in the grower's own words, see <http://www.leeks.com.au>.) A very similar situation occurred for strawberry crops in Victoria, where attempts to have no pest problems, in a crop where visual appearance is important, induced pest problems that were far worse.

Pesticides: problems and how they can fit with other management options

Our experience is that the less pesticide that is sprayed the better for the long-term control of pests. While this does not mean stopping all use of insecticides, it is important to remember that insecticides and miticides are poisons designed to kill certain animals. Therefore the attitude and practice of spraying an insecticide 'just in case' when there is no perceived target is likely to cause problems rather than fix them over the longer term. Similarly, the concept of using a cheap, broad-spectrum insecticide as a 'clean-up' spray at the end of the season to get rid of an unknown set of potential pest species is also not sound. Targeting specific insecticides against known pest species may be justified.

In a similar vein, after a pesticide that is known to be harmful to beneficial species has been sprayed it is often possible to find individuals of certain beneficial species that are still alive. This can lead some people to conclude that the spray was safe to that species. However, in an agricultural system where beneficial insects and mites are required to perform the important task of pest control, these beneficials need to be performing well: being active, feeding on the target pest and reproducing to produce on-going biocontrol agents. Individuals that survive a spray of pesticide that kills many or most of the other individuals in the same crop at the same time are unlikely to be performing at their peak! A human analogy would be a group of people who are being slowly poisoned with arsenic. Some die, others survive. Are the survivors likely to be performing as normal? So finding a live *Persimilis* (predatory mite) or ladybird in a crop following a spray of an insecticide or miticide does not mean it is safe.

It can be seen that the task of determining pesticide effects on beneficial species is more complex than counting how many individuals survive after 24 hours (see section on 'Safe insecticides'). In some cases there are lower rates recommended on pesticide labels where IPM is being used. However, although some *Persimilis* may survive when a lower rate of some miticides are applied, there will still be effects on other species of beneficials. This is something that is not discussed on a pesticide label, but the implication is that lower rates are suitable for IPM. Obviously the impact on all beneficial species needs to be considered if pest control is relying on a range of species. The decision on whether to apply an insecticide or miticide, and what to choose, needs to take into account the risk of economic damage from any pest following the application of the pesticide. Where pesticide-resistant pests occur, the assessment will also need to take into account whether or not the likely impact on the pest will be great enough to justify disrupting biological control agents for several pest species.

That decision is likely to be different for different growers, based on their perception of risk. We have seen examples where a grower has knowingly allowed

Figure 2.1   Pheromone trap and lure

aphid populations to build to such levels in hydroponic crops that the leaves and fruit were black with sooty mould and sticky with honeydew. This was because he was worried that the application of a pesticide to control aphids would kill the main predator of western flower thrips. However, the thrips were well under control, and would not have flared with one application of a selective aphicide. At an earlier stage in the life of the crop, when the thrips were not under control, the same pesticide application would have caused problems.

Other growers choose to react at the first sign of a pest, major or minor. In those cases the application of a pesticide to control a minor pest induces pest problems with the major pests, which then leads to further application of pesticides. This can occur even when no fully broad-spectrum insecticides are applied.

Pheromones

Many species of insects (particularly butterflies and moths) are known to find each other by smell, using chemicals called *pheromones*. In most cases they are pheromones that are released by females to attract males. This is one reason that male moths and butterflies have much larger antennae than females, as sensors on the antennae are used to detect the pheromones. In addition to these sex pheromones some species, such as European earwigs and *Carphilus* beetle, have aggregation pheromones to allow many individuals of the same species to gather in the same place.

The pheromones of many pest species have been synthesised and are now available in a controlled release form (in a rubber bung or tape lure) (see Figure 2.1). These are usually used as monitoring tools so that flights of pest moths or butterflies can be detected. The traps attract only the males, so the females are not trapped and are still able to lay eggs in the crop. The value is in early detection, not in control.

However, there are a couple of developments that have made the use of pheromones control methods rather than just a monitoring tool. The first is 'Mating Disruption' and the second is 'Attract-and-Kill'.

Mating disruption

Once the pheromone for a particular pest species has been identified, there is the potential to use it as a control measure, given certain conditions. The main condition is the size of the area being considered. The principle of using pheromones to cause mating disruption relies on the pheromones being able to attract all the males of the pest species in any given area. The males are attracted to the lures rather than to the real females, so the females remain unmated. That causes a lack of offspring (caterpillars), so a lack of pest damage. This strategy relies on the population of pests breeding within the treated area; if already fertilised females move in from outside, the strategy will fail.

If moths or butterflies fly in from outside already mated, then they do not need to mate again to produce caterpillars. So the area that can be protected using this method needs to be large: not only large in hectares but also a wide shape. If, for example, the minimum area that can be protected is 4 hectares, a square block with an area of 4 ha (200 m × 200 m) can be compared with a long thin block 50 m wide by 800 m long, because a square block has a reduced edge and more centre than the long thin block. Because of the 'edge effect', already mated females can fly in along any edge but may not move in very far. Such strategies are currently available for pests such as lightbrown apple moth in grapes and pome fruit, and codling moth in apples, for which there are commercially available lures and traps.

Attract-and-kill method

The attract-and-kill method uses the same pheromone and has the same limitations on the required effective area. The pheromone lures are placed on surfaces also containing an insecticide that will kill the male insects when they approach the lure. This may be within a typical bait, or may be in a mixture with the pheromone. In either case, the amount of insecticide used to kill the pest is extremely low compared to conventional spraying. The insecticide will kill the male moths or butterflies and so prevent successful mating, and hence prevent a damaging pest population from developing. A good example of a highly successful attract-and-kill approach is an aggregation pheromone used to control *Carpophilus* beetle in fruit and berry crops

Either method is likely to be effective in large-scale operations. For smaller-scale farms, where the edge effect is great, then the benefits of such an approach are poor.

Conclusion: minimal pesticide use

Pesticides that target certain species or groups of animal pests are, by definition, poisons that are regulated in most countries (but these regulations do not always support IPM, and this includes Australia). In addition to economic reasons, it makes sense that pesticide use be minimised, which is the goal of many government agencies around the world. Pesticides have a major place in agriculture at present, and there is no doubt that they have saved farmers from losing crops. However, they have also helped to create many of the pest problems with which farmers now have to deal and so the use of pesticides needs to be considered very carefully.

In this book we describe with the benefit of hindsight some of the problems with pesticides that were not anticipated when they were released, but there will certainly be other problems that are not predicted at present, or are disputed at present. The best that can be done is to minimise pesticide use and select options that are considered least disruptive to biological control. Most importantly if there is no target pest then there should be no application of a pesticide.

3

Pest species

In this chapter we discuss some of the different ways in which pests become problems, how they develop and where they come from.

Lifecycles and life stages

We sometimes hear from farmers that a pest was not present yesterday but caused significant damage the day after. Often this is not so, as the pest was simply present in lower numbers, and symptoms of damage had not yet been seen. Although sometimes pests can fly in overnight, many build up to damaging populations as successive generations develop.

The lifecycle of pests (and beneficials and all other animals) describes the main features of each life stage and the time each takes. The total time from egg to adult is called the *generation time*; that is, the time that must be involved in the progression from egg stage, through all juvenile stages to adult. Typically insects and mites begin life as an egg, then undergo a series of moults to develop through juvenile stages to complete development as an adult. Insects and mites have their skeleton on the outside. To grow they must shed this hard skin and form another. Each juvenile stage is called an *instar*, and the numbers of instars depends on the species. For example, many beetles often have just three instars, but black field crickets may have 11.

The lifecycle of the diamondback moth is an example of a caterpillar pest. There are male and female moths that must mate before the female will lay fertile eggs. The eggs are very pale yellow when they are laid, and darken just before they hatch. Tiny caterpillars hatch, then progress through several instars until they reach the pupal stage. This is a transitional stage from the grub-like caterpillar to the winged adult moth. In summer, this species may reach adult stage from an egg in under 3 weeks. In winter, development is extremely slow and can take months to pass through a generation, and development may stop when the pupal stage is reached.

A different type of lifecycle is where the insect or mite does not go through a pupal stage, but instead the juvenile stages look like small versions of the adult; for example, black field crickets. Each instar is just a little larger than the previous one, and if the adult has wings, the larger juveniles have wingbuds. The lifecycle is seasonal, and typically there is just one generation per year. Adults are active in summer and lay eggs in the soil. The eggs do not develop until temperature and moisture are suitable, so are triggered to hatch in autumn. They develop slowly over the winter months and turn into adults in late spring or early summer.

Some aphids, such as green peach aphid, can have a more complex double lifecycle, and populations can exist and reproduce without males. So green peach aphids populations can be all females, which give birth to live young, not eggs. Then the nymphs develop into adults without a pupal stage, but the adults can be either winged or wingless. The proportion of winged adults in the population at any time varies with species. Typically, if aphids find a good food plant that is young, they will produce wingless adults, as there is no requirement to disperse. However, if the plant they are on is senescing or the aphid population is extremely high, winged adults will be produced, as there is a need to find a new host. The fact that live young are produced and the generation time can be as little as 10 days means that populations can develop extremely rapidly.

Even the pests that have the shortest lifecycles still have a set of life stages that must be completed before the next generation is reached. So, for example, problems that develop with two-spotted mite over time are a result of the build-up of successive generations, and the consequent build-up of a massive population. The apparent sudden arrival of the pest is simply that the pest population has built up to damaging levels. The comparison may be made with insect pests that do arrive overnight in damaging levels. One example is Rutherglen bug. This pest actually does arrive overnight in massive numbers as winged adults. This can give farmers the false assumption that all pests occur in the same way.

There are very few examples of massive invasion of pests overnight. The development of a pest population may have been simply overlooked. This is often the case when monitoring for pests is based on symptoms of damage. For example, two-spotted mite causes silvering of leaves in a range of crops, including strawberries. When there are enough pest mites to cause damage symptoms, they have already increased from a low level to one that is significant. This event does not occur overnight. It can take weeks for the population of mites to reproduce and develop into larger populations. The smaller populations had not caused symptoms of damage, and therefore were not noticed ('they were not there'). The larger populations were suddenly noticed when they caused symptoms of damage ('they were not there yesterday'). Pest populations can be present at levels below the tolerance of farmers and indeed the detection of farmers. If the pest is present and given suitable conditions and the absence of beneficial species, the pest populations will increase and will become suddenly obvious.

This has led to the use of a *threshold* or count of pest numbers. Above the threshold number action is required, but below the threshold no action is required. This is different to the zero tolerance situation discussed in Chapter 2. In this case there are set thresholds for different pests in certain crops that have been established by researchers. Such thresholds can be found for many species, and a range of such can be found on the website of the Grains Research and Development Corporation (GRDC). It is better than the zero tolerance situation, which is unsustainable and undesirable if achieved by pesticide applications, but is still far short of being a good method of dealing with pests. We say this for several reasons. First, the threshold of pests where they will cause economic damage will depend on many different and unrelated variables. These include the age of the crop, the time until harvest, the end use, the market, the value of the crop, the life stages of the pest present, the number of beneficial species present and the type of beneficials present, the weather and the locality. To take all of these variables (plus more) into account will obviously require more than a simple count of pests per square metre.

For example, five large pest caterpillars per plant of any species in a susceptible crop of high value just before harvest with no beneficials present may be a serious problem, but 10 similar caterpillars in a young crop of low value with many beneficials present may be of no consequence at all.

As well as any set of thresholds determined by scientists, each grower will have his or her own threshold, which is the trigger for action. So one grower will do nothing at a certain pest pressure and another grower may be applying pesticides. Each will be acting on their own assessment of the level of pests that can be tolerated and cannot be tolerated.

Pest response when biological control is disrupted

It is very tempting for farmers to attempt to control pests with the application of pesticides. However, what is often not appreciated is the fact that pesticides can cause more problems than they create. This has been discussed in Chapters 1 and 2. We want to explore here the impact of pesticides on the predator–prey cycle and how it can have an impact on the control of key pests. We show here three graphs showing numbers of pests versus time in the crop. In Chapter 1 we gave three examples of the effects of removing biological control agents from strawberry crops. Here we describe what is common in many crops, not only the strawberry

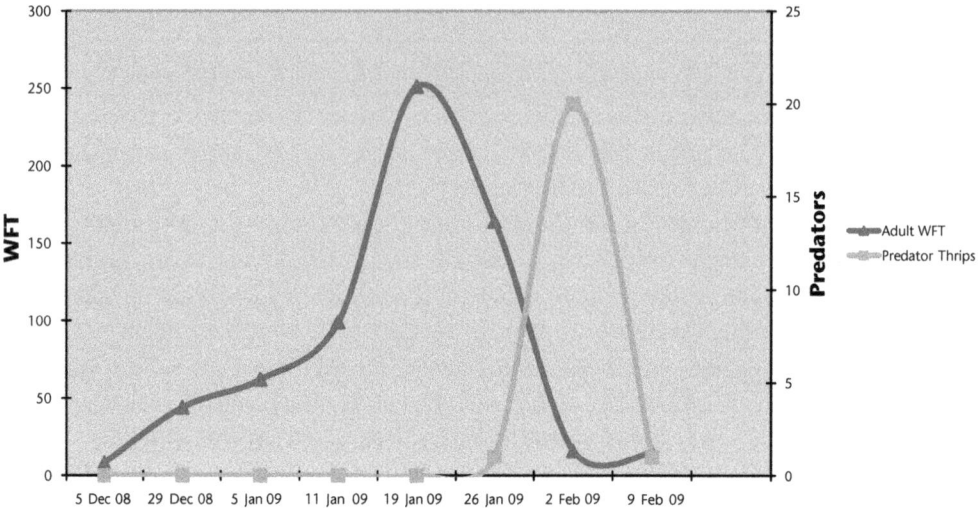

Figure 3.1 The classic predator–prey cycle. Data from a Victorian strawberry farm showing numbers of predatory tubular black thrips and western flower thrips (prey)

crops described earlier. Figure 3.1 is derived from real data while Figures 3.2 and 3.3 are generated from theoretical data to illustrate the point.

These three graphs illustrate the different ways in which predator and prey populations can develop with different management techniques. In Figure 3.1 we

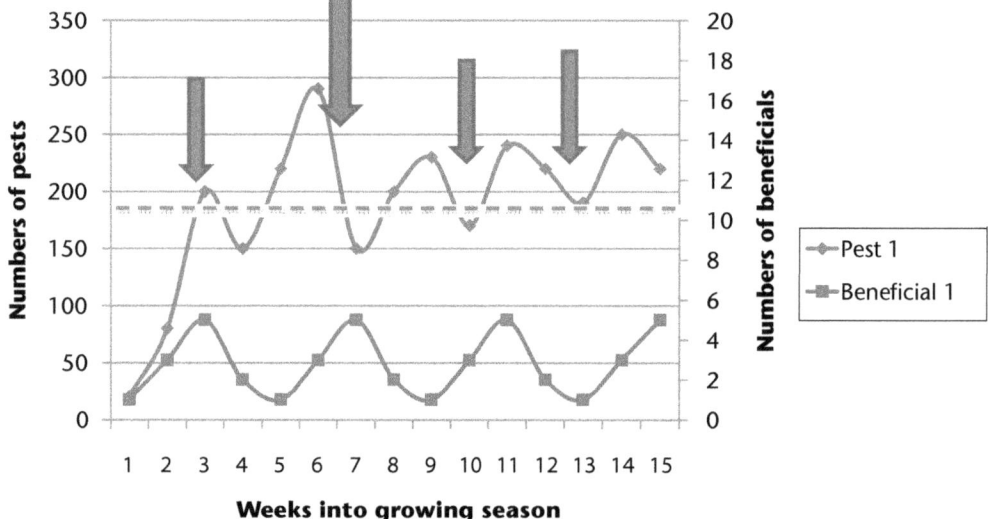

Figure 3.2 When pesticide applications (insecticides or miticides) disrupt the classic predator–prey cycle

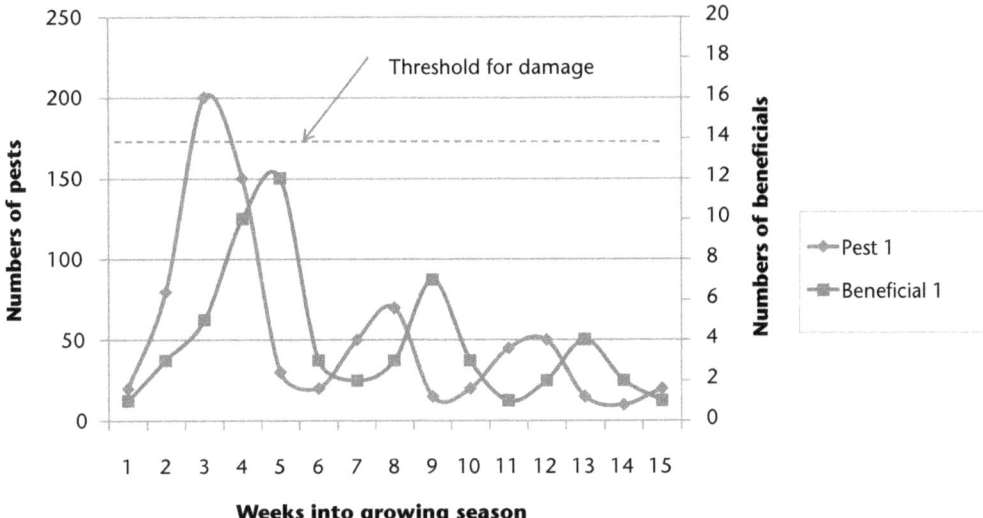

Figure 3.3   Predator–prey cycles within an IPM strategy

see a classical predator–prey interaction, using the same graph as in Chapter 1 (Figure 1.3). Prey numbers increase, followed by an increase in predators, leading to a population crash of first the prey and then the predator. Figure 3.2 illustrates what happens after the application of pesticides that are toxic to the predator without totally killing the population. When pest numbers reach a threshold (indicated by the dashed line) that the farmer does not accept, he or she applies a pesticide. In this example, it suppresses the pest population without eliminating it, and it also sets back the predator population. This means that the predator population never achieves control of the pest, although there are always predators present. However, in Figure 3.3 the predator population does gain control of the pest, then maintains control even though the numbers of predators are less than those in Figure 3.2 for most of the time. The aim is not always to maintain high numbers of predators but to achieve low numbers of pests.

When classical biological control is lost

Classical biological control typically deals with a pest that has escaped its natural enemies; for example, an insect or weed that has been introduced into a new country and has escaped its natural enemies. So the natural enemies in the home country of the pest are identified and brought in to deal with the pest. Usually a relatively small number of biological control agents are released, which are expected to increase by eating the pest, so classical biological control is achieved. Some very spectacular success stories have involved biological control of weeds in particular. These include control of prickly pear by the caterpillar *Cactoblastis cactorum* in Australia, control of cottony cushion scale in citrus by a ladybird

(*Vedalia* ladybird) in California, and control of water hyacinth weeds in Papua New Guinea and other countries by several species of weevils.

In all of these examples the biological control agents were restricted to that host, and could not move to other species. Therefore, the type of graph that describes the control is typically that shown in Figure 3.1, although the time scale may be in years rather than weeks.

Perhaps even more spectacular have been some of the failures of this type of control, notably the introduction of cane toads into Australia in an attempt to control cane grubs (various species of Melolonthinae), the introduction of foxes into Australia in attempt to control rabbits, and the introduction of a bird, Indian mynahs, to control cattle ticks. These were failures because, unlike the previous examples, the predators were generalists that were not restricted to one type of prey. These predators can happily survive and prosper on a range of other, non-pest prey so they themselves have become pests.

More recent introductions of insect biological control agents into Australia have been screened for host specificity, and if their host range or food preferences are not highly specific, they are not allowed to be introduced.

In this type of biological control the aim is restore a balance in the ecosystem so that an exotic pest again has its own natural enemies to help regulate the size of the population. Little management of the biological control agents is needed after the initial introduction into any area as the pest-beneficial populations maintain their own equilibrium. This can work well with some pests, but often this approach fails in agricultural systems where a range of pesticides is used. The pesticides interfere with the populations of beneficials, so the classical biological control is disrupted, as discussed above.

When existing biocontrol agents are disrupted

In addition to introducing specific biological control agents for specific pests, there is a range of naturally occurring species that can assist with control of pests, and often deal with them without further pesticide support. They may be either resident in the paddock (such as predatory mites or predatory carabid beetles) or they may be transient, following movements of pests (for example, lacewings, ladybirds and damsel bugs). These may be native or introduced species, as they are both established in the agricultural ecosystem.

Native generalist predators will often feed on introduced species of pests as well as on native species that may be pests. For example, the brown lacewing will feed on a range of introduced species of aphids, so populations of this predator can prosper in agricultural systems. It is a key biological control agent in lettuce crops in Australia and New Zealand where it eats lettuce aphid. The main factor disrupting the effectiveness of this predator in lettuce is the use of insecticides, including a seedling drench of imidacloprid. The seedling drench kills this predator by secondary poisoning, so it is not as obvious as acute effects of foliar sprays (Cole and Horne 2006), but is just as deadly to populations. Similarly, native predatory bdellid mites feed on lucerne flea, and will control them in most situations unless disrupted by pesticide applications. A range of native mites and predatory beetles prey on redlegged earth mite, and again, will control them in many situations unless pesticide applications kill the predators.

So whereas the attribute of being a general predator was a problem when importing biological control agents, it is actually an advantage when native predators that occur in agricultural ecosystems attack a range of species.

Where do pests come from?

It was mentioned earlier in this chapter how some pests (such as Rutherglen bugs) arrive in large numbers in a short period of time. Similarly, pests such as diamondback moth, *Heliothis* and plague thrips can also migrate over large distances on wind currents in Australia. Typically they disperse on hot northerly winds in late spring and summer. Many aphid species are active in spring and autumn, and have winged forms that disperse when the weather conditions and plant stages are suitable. However, it is not only pests that can disperse over large distances. Beneficial species can move in exactly the same way, usually following the movement of their prey or hosts. For example, brown lacewings and ladybird beetles (such as the transverse ladybird and spotted amber ladybird) also occur in large numbers from areas outside the crops at the same time that aphids are moving.

As populations of beneficial species are moving through agricultural areas, what happens when they encounter some areas where pesticides that would kill them are being used versus areas where no such products are used? Farmers often ask, 'What happens if my neighbour sprays harsh insecticides even if I do not?' The beneficial species will be killed, of course, in some areas, but this will not prevent them surviving when they land in areas where no disruptive pesticides are used. The main thing to remember is that the paddock becomes the habitat that is suitable for populations to develop. Transient species will find and develop populations in one paddock even if they do not survive in adjacent paddocks. It would be a better situation if large areas of agricultural land in a district were suitable habitats for transient beneficial species, as there would be a larger pool of beneficials, but it is not essential. The main concern is where pesticides disrupt biological control beyond the borders of where the pesticides are used. This is particularly true where the effect is sublethal, thus having an impact on the next generation not the immediate target.

Some pests can find a suitable habitat in weeds or crop residues, including volunteer plants from the previous crop. This provides them with a means of surviving without a particular crop being present, but immediately they can take advantage of (and possibly damage) a suitable crop when it is again available. Examples of this include vegetable weevil surviving on marshmallow weed, slugs surviving in crop trash or potato tuber moth surviving on self-sown potatoes.

4

Beneficial species (biological control agents)

In this book we talk a lot about beneficial species. Beneficial species are the natural enemies or biological control agents of pests; that is, the predators and parasites of the pests. These are called *beneficials* because they eat pests and so assist the farmer.

However, there are also a great range of beneficial species that are not directly involved in the control of pests. They may be involved in pollination or nutrient transfer (for example, bees and earthworms), therefore not only beneficial but essential to the functioning of any (agricultural) ecosystem. We would like to emphasise that there are great similarities between an agricultural and natural ecosystem in terms of the balance between plant feeding species and predatory and parasitic species, and the role that naturally occurring species have in controlling pests is very important. The difference between agricultural and natural communities has largely been a result of the impact of pesticide applications in agricultural systems. If we take the pesticides out of the system, then we can begin to see the role of naturally occurring beneficial species. The impact of pesticides (including soil applied pesticides) on beneficial species is therefore greatly underestimated.

Beneficial species can be predators or parasitoids of pests; that is, they eat them, either one pest per beneficial or multiple pests per beneficial. Predators may consume multiple prey while parasitoids may only consume a single individual. Examples of predators are ladybirds, lacewings, damsel bugs, bdellid mites and carabid beetles. Examples of parasitoids are wasps such as *Diadegma semiclausum* (parasitoid of diamondback moth) and *Aphidius* spp. (parasitoids of aphids) and also tachinid flies that parasitise caterpillars. Beneficial species can also be naturally occurring or commercially produced. Here we concentrate on the beneficial species that are available commercially.

Predators or parasitoids

The difference between predators and parasitoids is quite important, as they work in different ways. For example, there is a range of native species in Australia that eat aphids. Native species of ladybirds, lacewings, hoverflies and damsel bugs all eat aphids, and are not concerned whether or not the aphids are also native. In fact, populations of these native predators could actually be enhanced because of the presence of introduced species of aphids. There are also parasitoids (wasps in particular) that have also made it to Australia to parasitise some, but not all species of aphids. So, for the species of aphids that have escaped their parasitoids, the only biological control agents available are predators. This was the actual situation when lettuce aphid arrived in Australia in the 2003–2004 lettuce growing season. There are no parasitoids of lettuce aphid in Australia but there are predators, and these predators have been found to be effective in achieving control of lettuce aphid in commercial crops using IPM since its arrival (Horne and Page, unpublished data).

Predators such as hoverflies and ladybirds usually need a large population of aphids present before the adults of these predators will even lay eggs near them. So if there are only low levels of aphids, the predators are less important than the parasitoids. A female wasp of *Aphidius* that is looking for an aphid in which to lay an egg will happily lay one egg in the one or two aphids on offer, if that is all she can find. Each host will be sufficient to provide the necessary food and shelter for the offspring, and another wasp will emerge from each host. However, one or two aphids would definitely not be sufficient to sustain a ladybird, lacewing or hoverfly from egg to adult stage. So the relative impact of predators and parasitoids can depend on the population size of the pest (in this case, aphids). Parasitoids will be more important at low pest populations and predators more effective at higher pest populations. This effect can become even more significant if there are high levels of pests and the parasitoids are less able to find their hosts. Parasitoids often find their hosts by smell, so a fog of scent produced by a large number of potential hosts makes them less effective. However, at low pest levels they are extremely efficient at finding hosts.

Commercially produced beneficials

Trichogramma is a wasp that parasitises the eggs of its hosts (usually moth eggs). The difference between this parasitoid and most others is that it completes its lifecycle entirely within the host egg. That means that no caterpillar will eventuate if *Trichogramma* has stung an egg. The level of control that *Trichogramma* can achieve depends totally on timing. The wasps need to sting the host eggs before they hatch into caterpillars. Once hatched, the caterpillars are immune to *Trichogramma*. *Trichogramma*, which is also available commercially, is used to control pests that need to be controlled before they are larvae. (Examples include Macadamia nut borer; see the example in Chapter 8.) These wasps occur naturally, but can also be introduced artificially to elevate naturally occurring populations.

Other wasps parasitise the larval stages (caterpillars) rather that the eggs, and there are even species that sting the eggs but do not kill the caterpillar until it has moulted several times. These species are only apparent at the time that they kill the host, and this is usually only in the pupal stage. Hormones trigger the pest to pupate. These hormones also probably trigger a parasite living within the host to take over and kill the host. So the parasites of the pest are tied closely to the lifecycle of the pest. If the weather is cold and the generation time of the host caterpillar is extended, then the generation times of the parasitoids and predators linked to this pest are also extended in exactly the same proportions.

Many beneficial species have been developed commercially because of the requirements of glasshouse vegetable and flower producers. In European and many other countries (including New Zealand, but not Australia) the need to have a range of biological control agents for pests has been because of the need to protect bumble bees. Bumble bees are used as pollinators, thus saving on a great amount of labour costs, and in no small way have been responsible for the development of integrated pest management (IPM) in many countries. These bees are not available in Australia, so the development of IPM in Australia has been a result of other origins. The fact that IPM in Australian glasshouse crops is well established is a testament to how effective IPM can be, even given the lack of bumble bees to drive the system.

Commercially available beneficial species also allow the grower to decide to use a biological-based approach instead of a pesticide-based approach simply by buying a different product. The grower can choose to buy some biological control agents for a particular pest instead of relying on a pesticide or on naturally occurring beneficial species.

Examples of commercially available beneficial species

Full descriptions of the range of beneficial species that are available can be found on websites of the producers themselves (for example, those of two large companies, Koppert and Biobest in Europe and USA) or of associations, such as Australasian Biological Control Inc. Two examples of beneficials that are commercially available in many countries are *Persimilis* (for control of two-spotted mite) and *Encarsia* (for control of greenhouse whitefly).

Two-spotted mite and *Persimilis*

Control of two-spotted mite by the predatory mite *Persimilis* is one of the key elements of IPM in a range of crops. Two-spotted mite is a particular problem for flower crops (such as roses and gerberas) and is resistant to many of the miticides that are available. The dense canopy of leaves makes contacting the pest with miticides very difficult, so dealing with it is often the main pest management problem for these flower growers. However, two-spotted mite is also a major concern in other crops such as apples, pears and strawberries, and can even be a problem in grapevines and tomatoes.

Persimilis is a predatory mite that eats two-spotted mite and does not survive without it. Commercial producers grow the mite on plants infested with two-spotted mite, then either extract the predators and send them out to farmers or else harvest the leaves when *Persimilis* has eaten practically all of the pest mites. *Persimilis* mites eat all stages of the pest (eggs, nymphs and adults), and when they run out of two-spotted mites to eat they will resort to eating *Persimilis* eggs.

Persimilis is the key predator in many crops, such as strawberries, because if it is not established and in control of two-spotted mite, the farmer will have little hope of getting other biological control agents (as for whitefly or aphids) to work. That is why, in many horticultural crops, control of two-spotted mite is the key to controlling a range of other pests.

Whitefly and *Encarsia*

Control of greenhouse whitefly with the wasp *Encarsia* is a very well-established method. Commercial growers of crops affected by greenhouse whitefly (for example tomatoes, capsicums and many flowers) can buy and release *Encarsia*. The wasps are sold as pupae that will emerge after the grower has received them, and these wasps will seek out and sting whitefly nymphs of a suitable age. When they sting the whitefly, the wasps are actually placing an egg inside the host insect. The egg develops into a maggot, which eventually kills the whitefly larva, and instead of another whitefly adult emerging, an *Encarsia* wasp emerges.

By introducing *Encarsia* and *Persimilis* at an early stage in the life of the crop, a farmer can establish a breeding population of the wasps or predatory mites which can give on-going control of the pests. In this way they are used to 'inoculate' a crop before they may otherwise occur naturally. It is sometimes also possible to 'inundate' a crop with a large number of beneficial individuals (for example, the wasp *Trichogramma* can be released in massive numbers when desired). However, it is usually a better option for farmers to look at fostering populations of resident beneficials rather than using them like a pesticide.

Lifecycles

Beneficial species can have very different lengths of lifecycles. For example, ladybird beetles begin life as eggs, hatch into small larvae and, depending on temperature, take about 3 weeks and several moults to reach the adult beetle stage that we all recognise. The adult beetle may then stay alive for a year or more, producing eggs when conditions are suitable. However, not all beneficials have the same timing. Parasitoid wasps that attack aphids would complete their lifecycle in less than two weeks in summer. Carabid beetles that help control slugs often have a 1- or 2-year development time as juveniles, so they are much slower to respond to good conditions but much more vulnerable to adverse conditions.

The lifecycles of parasitoid wasps are usually linked to life stages of their hosts. So when the larval parasitoid is a maggot living inside its host, its development may speed up with higher temperatures and slow with lower temperatures, just as the host's development changes. That means the life stages of both parasitoid and host will develop at the same rate.

Movement of beneficial species

There are species of beneficials that are resident in a paddock, such as predatory carabid beetles and predatory mites. Then there are the transient species that move through any district with the seasons, such as green and also brown lacewings. Green lacewings are important in the control of key pests of vineyards (lightbrown apple moth and long-tailed mealy bug) in southern Australia. Our observations over many years suggest that in some years there may be a short time when the green lacewing adults are flying into vineyards, so there is a critical time when these individuals can colonise. If they are killed by even a relatively non-persistent insecticide application and there are not subsequent flights, the opportunity to establish this species of beneficial is lost for the entire season. Similar situations may occur with species such as damsel bugs (predator of caterpillars and other soft-bodied insects) and brown lacewings, both of which are important in many vegetable and broad-acre crops in southern Australia.

Naturally occurring beneficial species

In most crops in which we work it is the naturally occurring species of beneficials that are responsible for the bulk of the pest control. These are incredibly abundant and reliable because they follow populations of pests. In most of our work we rely on naturally occurring populations of beneficials, not on commercially produced individuals. The main thing for farmers to do is not to kill them! This seems very simplistic, but is actually the basis for good biocontrol in most agricultural ecosystems.

This last recommendation is not a flippant remark. There are vast numbers of beneficial species that move from one vegetation type (native or introduced) to another each year, following populations of prey. The impact that these native predators can have on populations of pests is massive and the main disruption to that is the application of broad-spectrum insecticides. It is a much better situation if the species of pests and beneficials are known so that any application of pesticide can be targeted to have minimal impact on the key predator or parasitoid in any given crop.

There is a range of beneficial species that can be utilised by farmers, depending on the crop being grown. Parasitoid wasps that attack caterpillars and aphids can be very important, but so also can be predators such as ladybirds, hoverflies, damsel bugs and other predatory bugs. The relative impact of each of these species needs to be measured against the impact of any pesticide against any particular pest.

Common questions about beneficial species

Here we list just a few of the most common questions that we are asked about beneficial species, along with the answers to those questions. In all cases, the answers need to be considered in relation to *populations* of beneficials rather than just individuals that may be present on any one day (see Chapter 1).

How many different beneficial species do I need?

It makes sense to utilise all of the beneficial species that are available, although usually one or two species often account for most of the biological control achieved. However, in other situations, such as control of redlegged earth mite, there is a complex of different species that all contribute to control by predating on different life stages. Which species are dominant will depend on such factors as location, crop type, weather and pesticide history. Therefore local observation is most important in determining the key species in any given situation.

How do I attract beneficial species into my crop?

Beneficial species are attracted by their prey, but some other factors can help to sustain them. The most important factor is not to kill them, and then to create the right environment in which they can survive.

Why can't I find beneficial species?

High numbers of predators or parasites of pests will only be found if there is an abundance of food (pests) for them. If there is no food, there is nothing to sustain them (see Predator–prey cycle' in Chapter 1). This is particularly so if the predator or parasite is restricted to one or very few species of prey. So, for example, you will only find many hoverfly maggots where there are many aphids, and you will only see many wasp parasitoids, such as *Diadegma*, where there are suitable host caterpillars. If there is an abundance of hosts or prey but no beneficial species, another cause, such as pesticide residues, may exist. Not seeing a lot of beneficials may actually be a good sign; that is, well-balanced biological control is happening in the background.

Where can I buy beneficial species?

There are Australian and worldwide companies that produce beneficial insects and mites for particular pests (and particular crops). Protected cropping (glasshouse and polyhouse crops) are particularly well serviced by these commercial insectaries. However, a common mistake is to attempt to only use these beneficial as a type of 'biological insecticide' and expect instant results. Again we refer to the 'Predator–prey cycle' section in Chapter 1, and suggest that in most situations the beneficials released are simply an inoculation to begin or to boost a predator population so that it can eventually gain control over a pest population.

For information on the beneficial species that are available in Australia and New Zealand, refer to <http://www.goodbugs.org.au>. In other parts of the world, look up sites of producers of beneficial insects and mites, such as <http://www.koppert.com>.

Effectiveness of biological control

In many crops and in relation to the control of pests, the impact of naturally occurring biological control agents can be massive. Control of lettuce aphid in Australian lettuce crops could rely totally on naturally occurring biological control, and needs no assistance from pesticides. Similarly, control of pests such as greenhouse whitefly can be achieved with spectacular success by releases of the wasp *Encarsia formosa*. What is needed is to ensure that the biological control is not disrupted by pesticides applied for other, usually less serious pests. The impact of pesticides applied for other pests (including caterpillars and aphids) is often the reason that biological control alone is disrupted and not allowed to deliver control of key pests. The role of biological control is usually greatly underestimated, and it needs to be allowed to work simply by removing the pesticides that kill them. Integrating biological control with other control options is the key to a total change in pest management.

5

Cultural controls

What are cultural controls?

We define cultural controls here as any physical or management method or decision that is used to influence either pest or beneficial species, or their impact on the crop being grown. The term 'crop' here means whatever the farmer is trying to produce, whether pasture, roses or broccoli, for example. That means the aim is not always to kill pests or build up levels of beneficial species, but simply to use management to manipulate the impact of pests. One of the most visible examples of a type of cultural control is where grapes are grown with the use of netting to prevent birds from damaging the ripening berries. Simply, a physical barrier is placed between the pest and the crop. Netting with a finer mesh is sometimes used as a means of preventing moths from accessing particular vegetable crops.

One of the oldest cultural control methods that is still available to growers, and which is a very powerful means of reducing pest damage, is rotation of crops in a paddock; that is, not growing the same crop for successive years in the same ground. This helps reduce pest populations that rely on a particular crop or crop type to survive, so pest pressure is reduced. Conversely, if successive crops of the same type are grown in the same ground, pest pressure will carry over and build up from crop to crop. A problem with many types of cultural controls is that they may not give immediate results, so the benefits can often be overlooked.

In some situations the cultural options are more effective than the biological or chemical options that may be used against the same pest. One example is the use of soil management and irrigation to control potato tuber moth in potatoes. A second

Figure 5.1   Nets used as a physical barrier

example is the timing of planting to control slug problems in canola and aphids (and virus) in cereals. In most other situations there are contributory cultural controls that do not necessarily offer the most important control option. A third example is the use of resistant varieties of plants, such as lettuce resistant to lettuce aphid.

Example 1: Soil and irrigation management in potato crops

Potato tuber moth is an important pest in many solanaceous crops, including potatoes. The main damage that is of concern to potato growers is caterpillars burrowing into tubers. Although the caterpillars can cause damage to leaves, it is usually the damage to tubers that must be prevented. Given that the tubers grow below the soil surface and that the adult moths and the caterpillars that they give rise to are present at least initially above the soil surface, there is the opportunity to prevent access to the tubers. Preventing such access depends on there being an intact layer of soil above the developing tubers that caterpillars cannot penetrate. They are able to penetrate the soil only if it is cloddy, cracked or so shallow that tubers are exposed. The cultural controls in this case rely on preventing such conditions by good soil preparation before planting, preparing a large 'hill' and preventing the soil from cracking or filling in developing cracks by the use of appropriate irrigation. All of these measures work to achieve the same end: to

Figure 5.2 Broccoli after broccoli: a poor use of rotation of crops

provide an intact layer of fine soil over the developing tubers that forms a barrier to potato tuber moth caterpillars of all stages that are present above the soil surface. In the same way that nets on vineyards form a physical barrier between grapes and birds, soil is used to make a physical barrier between potato tubers and caterpillars.

Example 2: Planting time in canola and cereals

Pests such as slugs can severely damage canola plants and other broad-leaf crops at the cotyledon stage and prevent the crop from establishing (and so are often referred to as *establishment pests*). Unlike cereal crops, germinating canola produces cotyledons from the seed before the first true leaves develop. If the cotyledons are destroyed before the first true leaves develop, the plant cannot develop the true leaves, so the plant dies. Once true leaves have developed, the plant is much less vulnerable to damage by establishment pests.

So the time that the plants spend in the vulnerable cotyledon stage can strongly influence the degree of damage by a set number of pests per square metre. If the crop is planted relatively early, the plant will progress rapidly through the cotyledon stage to the true leaf stage, so the pests such as slugs have only a relatively short time to cause damage. However, if the crop is planted later when the weather is colder, the plants will spend a relatively much greater time in the vulnerable stage, so the same density of pests will have time to cause much greater

Figure 5.3   Potato tuber damaged by potato tuber moth

damage. A very powerful tool to avoiding damage by establishment pests such as slugs is to select the time of planting.

An almost opposite method, but still using time of planting, can be used to minimise the damage caused by pests such as aphids, vectoring barley yellow dwarf virus (BYDV) in cereals. Aphid flights are typically seasonal, especially in southern Australia, and so if it is possible to avoid planting when aphids are highly active, it is also possible to avoid the problems that aphids cause. For example, planting before mid-May in southern Victoria will usually expose young seedlings to flights of aphids. However, planting after mid-May will usually avoid exposing young plants to damage from large flights of aphids. Therefore the requirement to use insecticide treatments, even seed dressings, is very much reduced.

Variety selection can be the most powerful control measure for some pests. For example, if you plant varieties of lettuce resistant to lettuce aphid, no insecticides are needed for this particular pest. There are many other examples are plants that are resistant to particular pests or the diseases that they vector; for example, varieties of wheat resistant to barley yellow dwarf virus; lettuce resistant to lettuce aphid; lucerne resistant to aphids.

Genetically modified (GM) crops could be considered in this section, as they are particular varieties, albeit created by a very different method than conventional breeding. For example, transgenic cotton with BT toxin (*Ingard* cotton) is resistant to caterpillars. From the farmer's point of view, in the options available to control pests, resistant varieties (including GM varieties) are a significant tool that can be used. However, if they are used as a stand-alone or 'silver-bullet' type of control, rather than as a component of an IPM strategy, their value will not be realised.

Figure 5.4   The cotyledon stage of canola showing damage by lucerne flea

Figure 5.5   The true-leaf stage of canola

Although plants have natural defences against insect attack, these often take the form of distasteful chemicals. Over the centuries, in our selection of varieties of food crops we have in many cases reduced the levels of these chemicals because we want our food to not taste bitter. So brassica weeds such as wild radish are less palatable to pests such as diamondback moth than a commercial cabbage, just as it is to us.

Example 3: Sequential planting

There are particular advantages and disadvantages in *sequential plantings*, where the next planting is made immediately adjacent to the previous planting so that there is a range of ages of plantings within any one paddock. Each planting is made beside one older planting and before the next younger planting. It is easy to imagine that the arrival of a pest and the subsequent arrival of a range of beneficial species, as described earlier in this book (the section on 'Predator–prey cycle' in Chapter 1) will happen with each planting. But the greatest fluctuation of pest and beneficial numbers will be in the first planting. Successive plantings will benefit from the presence of beneficial species in adjacent crops, but the first planting in a paddock will not get this benefit, so the problems will be worse here. The first crop planted in a new paddock can be expected to have the worst problems with pests such as aphids. As beneficial insects build up, pest problems will be less with each planting.

Manipulating non-crop plantings for pest management

There are several types of plantings that have been used with the primary aim of pest management. These include planting strips of non-crop plants (such as grass, lucerne and flowers) and management of inter-row cover crops (plants placed between vine or citrus plants). They can also include planting trap crops; that is, a non-crop plant that is more attractive to a particular pest than the crop.

Each of these methods involves utilising details of either pest or beneficial species behaviour and biology either to encourage beneficial species or make the situation less favourable to pests.

Mowing every second row (rather than every row) in a vineyard or orchard at any time can result in flowering grasses. That means that a pollen source (grass pollen) is present to provide an alternative food for predatory mites to supplement their diet of invertebrate prey. Therefore, a larger population of predators is carried than would otherwise exist.

Example 4: Strip plantings

Planting strips of non-crop plants can include planting flowers that provide a nectar source for beneficial species but not pests. There is limited information on

Figure 5.6 Mowing alternate rows in a vineyard

Figure 5.7 Grassy strips to assist with biological control in lettuce

this approach, as there is a need to avoid providing a nectar source for pests. In both cases the provision of a supplementary food supply such as nectar or pollen can result in a greater number of offspring from individuals that feed on these plants. The type of flowers present and the mouth parts of insects will determine which species can access the potential food supply and which cannot.

Another approach is to plant strips of plants, such as cereals, that will attract a pest, such as aphids, but not a pest of the crop being grown. In this example, the aphids that feed on cereals will not attack vegetable crops such as lettuce or broccoli. However, generalist predators such as brown lacewings will feed on cereal aphids and the aphids that attack vegetables. In this way, strip plantings of grasses at the time of year that cereal aphids are active will encourage populations of brown lacewings to become established in the strips. If the grasses are planted ahead of the vegetable crops, populations of aphid predators will become established even before the vegetable crop is planted, thus avoiding the typical lag that occurs in the predator–prey cycle (see Figure 3.1).

Example 5: Border plantings

It is very common for us to be asked, 'What can I plant around the crop to encourage beneficial species?' A range of native plants with different flowering times and different heights is usually a good option. Australian native plants are often not so attractive to exotic pests, and the different flowering times means that a pollen and nectar source is available for extended periods. The different height of plants means a more complex understorey is available, providing more habitats for a range of species.

Border plantings are good for a range of reasons, but pest management is probably one of the least important. It is not likely that border plantings will affect populations of predatory invertebrates for more than a short distance from the border. Insects that like to live in trees will prefer trees to a low-growing crop or pasture. Although there may be some benefits, they are going to be restricted. Parasitoid wasps that fly may be more favoured if they can feed on the nectar from the border planting. In both cases, most effort needs to be on changing the environment within the paddock to suit beneficial species and being careful to manage these well.

Controlling weeds

Pests that attack brassicas are often restricted to brassicas. This can be used to help control or reduce the impact of pests such as cabbage white butterfly and diamondback moth. Pest populations can only develop on brassicas, so controlling brassica weeds is important. In particular, minimising access by the two pests mentioned to flowering brassicas will reduce the number of eggs that female moths and butterflies will lay in the crop, thus reducing the pest pressure. Conversely, allowing the moths and butterflies access to nearby flowering brassicas (such as around the perimeter of the crop) will mean a greater pressure from these pests, particularly close to the flowering plants, as more eggs will be laid.

Weeds can also be a reservoir of both pests and diseases. For example, self-sown (*volunteer*) potato plants can carry both the viral diseases of potatoes and the insect vectors that transmit them. So controlling these particular weeds is of great importance to potato growers, and particularly seed potato growers. Self-sown potatoes tend to emerge and flower before the main crop plants. The flowers are attractive to thrips, which vector tomato spotted wilt virus (TSWV), so self-sown plants that carry this virus are a serious problem for the health of the crop. Aphids could easily move from self-sown plants to the main crop, and many other viral diseases of potatoes are vectored by aphids.

Controlling cape weed is an important part of a strategy to control redlegged earth mite. The mite prospers on this weed, so pest mite populations can increase massively if cape weed is abundant. Redlegged earth mite prefers broadleaf plants, so in a paddock where wheat or barley or pasture grasses are free of broadleaf weeds the pest does poorly. Conversely, if there are large areas of weeds such as cape weed, the pest mites will become far more abundant. As both the weed and the pest mites are dormant over the warmer months, problems will be seen in the autumn following any weed problems in the winter.

Plant health

Most farmers, supported by scientific studies, agree that stressed plants are more likely to be attacked by pests such as aphids. Conversely, growing a healthy crop (which is desirable for many reasons) is going to assist with pest management. Is a healthy crop immune from pests? We suggest not, partly because the varieties of plants that we have bred have been selected for some characters often at the expense of defensive chemicals, and partly because we have not had the opportunity yet to test this assertion. Many organic farmers would disagree, and aim to grow plants with a Brix level (a measure of plant health) over 14, as it has been claimed that plants with this degree of health will repel pests. Whether this is true or not, a healthy plant is obviously more desirable than a stressed plant, and if that level cannot be achieved, or pests are still a problem, other pest management options are still available.

Onion crops and mulches

Tasmanian farmers produce a significant proportion of the Australian crop of onion, and most of this crop is produced for export. That export may be to other states of Australia or other countries. One Tasmanian farming company that produces onions in recent years has adopted an IPM approach to all of its production. The onion production component of its business has been focused on control of onion thrips, both during the production phase (in the paddock) and after harvest (in bins).

Before the adoption of IPM by the company, the methods used in-field were pesticide based, aimed at killing all thrips present in the growing onions. The effectiveness of such a strategy is dubious, but is currently mainstream procedure in Australia and New Zealand at least. A changed strategy needed to be trialled before being more widely implemented, and the basis for the new strategy needed to be understood.

The theory that we suggested was that there were predators of onion thrips that could eat onion thrips. These could occur naturally, but could also be encouraged. After harvest there was the option of using commercially produced thrips predators (mites). So the strategy that we proposed was to encourage the predators of thrips in the field and to release commercially produced predators post harvest.

The complexity of the invertebrate population at the soil-surface level was increased by adding mulches. This attracted flies (fungus gnats/*Sciaridae*) which in turn increased the size of the population of soil-dwelling predatory mites such as those in the genus *Parasitus* (no common name). When the mites had eaten all of the fungus gnats, they were hungry, so began to feed on the only other available food source: thrips. The trials that the company conducted convinced them to try this approach on a bigger scale. It worked, and it is now standard practice. No insecticides are needed to control onion thrips in onions.

If any post-harvest actions are required, predatory mites can be released into bins of onions rather than pesticides or fumigants being applied. This deals with crops where control of thrips has not been adequate before the harvest, whatever the approach taken. The Tasmanian company has proved to itself that a reliance on pesticides is not necessary or even reliable to control thrips in onions. Some thrips species could be considered pesticide-induced pests, so that 'the more farmers spray then the more they need to spray'!

Crop hygiene

When the same, or similar crop, is to be grown in the same area, there is the risk of a build-up of pests and diseases. Diseases that are specific to one type of plant can be controlled to a large extent by minimising the carry-over of plant material from one crop to the next. In many crops this will rely heavily on rotation and controlling self-sown (volunteer) plants, so that there is no suitable host for the pest or disease to feed on. This is an important aspect of some certified seed production schemes, for example potatoes, where no potato crops are allowed to be produced in the same ground for up to 7 years in order to minimise pest and disease problems.

For permanent crops (such as trees and vines) there is no opportunity to rotate crops in the ground, but hygiene is still very important. Removing rotting fruit is one means of minimising the build-up of pests such as vinegar flies and *Carpophilus* beetle, and also preventing the build-up of high levels of specific pest populations that directly damage the crop; for example, cocoa pod borer in cocoa, and codling moth in apples. However, the permanent crops can develop a high level of biodiversity of beneficial species, which is of great value.

Cocoa pod borer

Cocoa has been an important crop in several different countries in Africa and South-east Asia over the last few decades. In Malaysia first, then in Indonesia, the crop has been initially highly profitable but the production has then declined. One of the greatest reasons for the decline part of the sequence is an insect pest: cocoa pod borer. It has been estimated (International Cocoa Organization) that losses by the year 2000 were around US$40 million. In the early 1800s it was responsible for the decline of the industry in North Sulawesi.

The moth lays eggs directly onto the cocoa pod, and the caterpillar that hatches out burrows into the pod and can also cause unevenness and premature ripening. Control of this pest depends largely on removing trash cocoa pods – the removal of pods that are not harvested – even though other control measures are available and should be used in an IPM approach. Even if they are not harvested to be sold the pods should be collected and removed from the farm. This is in order to break the lifecycle of the pest and remove a significant proportion of the pest population. Removing the trash is a job that is needed for control of the pest, but is something that does not give an immediate cash return on the labour cost expended, so is a job that is often not done. The price of not removing the trash is not seen immediately, but in the longer term can cause total failure of the industry.

6

Integrating control measures to maximise degree of control

Integrating the three control measures available

The previous chapters have outlined the only three control measures available to farmers wanting to protect their crops from invertebrate pests. Farmers can use any of the *biological, cultural* and *chemical* control options that are available, but there are no other additional options that we can see. The use of genetically modified crops is simply the use of another variety, and any new pesticides are simply new chemical options. New biological control agents are possible, but not likely for any given crop. So what can a farmer do as the best option to control pests?

The best option for any farmer of any crop or pasture is to look at the possible control options to try and build a strategy that is based on biological and cultural controls and needs support only when necessary from chemical applications. This is our definition of integrated pest management (IPM). Integrated Pest Management uses all available tools at any given time, so is not a fixed recipe for pest control but rather an approach to the best way to achieve pest management. A method of developing an IPM strategy for *any* crop is to complete the blank table below. (Note again that an IPM strategy is developed for a crop, not for a particular pest.)

In this example it is necessary to identify any possible biological control agents for each pest: there may be some highly effective beneficials or there may be none. The next step is to identify the available management options either to discourage

Table 6.1 IPM in (crop specified)

Pest	Beneficial (biological control agents)	Cultural (management)	Chemical
1			
2			
3			
4			
5			
6			

the pests or encourage the beneficials (or both). The final step is to decide on the available chemical support options that are compatible with all other control measures, should the biological and cultural controls not be enough on their own. There may not always be a totally safe option, but all pesticide options need to be considered, and the effect of any application (on beneficials) needs to be known before it is applied.

The role of monitoring

The reliance on biological control agents and the support tools of selective pesticides means that correct identification of both pest and beneficial species is necessary. It is also necessary to have assessments made regularly, or at least during important stages in the life of the crop. Crop monitoring is made with the aim of allowing timely decisions to be made and that these decisions take into account the degree of biological control that can be expected within the IPM strategy.

Manuals are available for many crops, describing the many different species that can be present (pest and beneficial) within each crop. These can be daunting to anyone who is contemplating changing from a pesticide-based approach to IPM, as it may seem necessary to know everything about each species before starting. However, this is not so. It is better, in our view, to start monitoring, then look for different or new aspects and methods, then seek assistance with identification as necessary.

Recording information is useful, especially for looking at historical trends and activity of species at times of the year, but the primary aim is not to produce records but to allow decisions to be made.

Traps and lures

A range of tools can be used to help with monitoring, and in particular several different types of traps. The type of traps used and the intensity of monitoring needs to be appropriate for the crop type, the species of invertebrates being monitored and the degree of risk or nervousness incurred by the farmer or monitor. Pitfall traps can be used to measure the types of species present that are active on the soil surface. Refuge or shelter traps can be used at certain times of year to assess activity of species such as slugs, carabid beetles and earwigs. Yellow sticky traps can be used to see what is flying in or above the crop, and pheromone traps (see Figure 2.1) can be extremely useful in monitoring the activity of particular pest species. However, one of the most useful monitoring method is simply direct searching.

In almost all of the above examples of trapping (the exception is pheromone trapping) there remains a requirement for the person doing the monitoring to have the skills necessary to be able to accurately identify the species present. As mentioned above, this requirement can be daunting but should not be so. There are guides and support tools available for those who seek them out.

Making changes in pest management

The benefits of integrating the use of biological, cultural and chemical control options in a compatible way (that is, Integrated Pest Management) is to us the logical approach to dealing with invertebrate pests. It is promoted by many government agencies around the world, and the benefits are many (see Table 6.2), but adoption rates are often slow (Table 6.3) and remain low for most agricultural sectors (Horne *et al.* 2008). Despite the methods required for adoption of IPM (and promoting other changes) being established, there remains a reluctance to acknowledge and address the issues required to use IPM.

Our own experience working within our company, IPM Technologies Pty Ltd, is that we have had great success in helping farmers to deal with a range of pests by adopting an IPM approach. These areas of success include insecticide-resistant pests, such as western flower thrips, two-spotted mites and lettuce aphid; dealing

Table 6.2 Advantages and disadvantages of adopting IPM

Advantages	Disadvantages
Reduced dependence on pesticides	More complex than control by pesticide alone, and requires a shift in understanding
Increased safety to farm workers, spray operators and the community	Requires a greater understanding of the interactions between pests and beneficials
A slower development of resistance to pesticides	Requires a greater understanding of the effects of chemicals
Reduced contamination of food and the environment	Increased time and resources
Improved crop biodiversity	Level of damage to the crop may initially increase during transition to an IPM program, in some horticultural crops

Table 6.3 A comparison of IPM and chemical-based supports (derived from Bajwa and Kogan 2003)

Pesticides	IPM
Compact technology	Diffuse technology with multiple components
Easily incorporated into regular farming operations	At times difficult to reconcile with normal farming operations
Promoted by private sector	Promoted by public sector
Aggressive sales promotion supported by professionally developed advertising campaigns	Promoted by extension personnel usually trained as educators not as salespersons
Results of applications usually immediately apparent	Benefits often not apparent in the short term
Consequently: pesticide technology was rapidly adopted	**Consequently: Adoption of IPM technology has been slow**

with pests of crops grown for cosmetic appearance or for which their appearance is critical; and insects that are vectors of viral diseases (such as barley yellow dwarf virus in cereals, tomato-spotted wilt virus in capsicums and potatoes, and leaf-roll virus in potatoes). The reason for our success is that we have largely operated in the commercial sector and so have used methods more in the left-hand column of Table 6.3 rather than the right-hand column.

Table 6.3 highlights the differences that may be experienced from a farmer's point of view when deciding between a pesticide-based strategy and an IPM strategy. Furthermore, for any individual farmer or farm a pesticide-based strategy is known, legal, and it works, which makes the decision to adopt something unfamiliar and unproven (on the farmer's own crop) even more difficult.

As the table shows, the results of IPM may not be seen in the short term. This is true not just of IPM but of other pest management decisions or changed practices, which might incur a cost in the future in either pest management or other issues. For example, there could be a choice between using one or another pesticide. One may give control of a pest now but give rise to problems with control of pests in the future. The benefits are seen in the life of the crop but the problems are not. Similarly, particular practices such as rotation may give substantial benefits in the longer term but these benefits will probably not be seen in the life of the current crop. Where the crop is a fairly short-term crop, such as many vegetables, the benefits need to be seen within a matter of days or weeks if they are to be evident in the life (and costings) of the current crop. Problems may be seen only several crops later, so are more difficult to cost directly within only one crop.

Decision-making

For every action made that influences pests or beneficial species someone must make a decision to take that action. Sometimes it is based on poor information or is just a standard practice, but there still must be a decision made. In this book we have tried to show the range of options available to control pests and the information needed on which to base a decision. It is not always easy, and different farmers will make different decisions when given the same set of pest numbers and crop types. This is understandable; the main goal should be that the consequences of any actions are understood beforehand rather than only discovered afterwards. For example, an application of an insecticide may kill some species of beneficial insects. The farmer may decide to do this anyway after weighing up the relative benefits of the application in a certain time period versus the loss of the beneficials for a given time.

We often hear farmers justify the application of a pesticide by saying, 'Well, we had to.' It is more accurate to say 'We chose to' apply the pesticide, as there are almost always several options to choose from, although some of the options should have been adopted weeks or months or more before. When the first possible options are not used, the number of options available obviously are reduced. This highlights the difficulty in making a change to a different type of pest management and often having to wait for the often slower effects of biological and cultural controls to become evident. Very often the hardest decision for a farmer used to spraying insecticides is to 'do nothing' and wait for biological control agents to gain control. Similarly, it is also extremely difficult for an adviser (who may be a chemical reseller) who is used to recommending pesticides as the basis for control to recommend not to spray an insecticide, but instead wait for biological control agents to work. In particular, it is difficult when a farmer says, 'I have a problem with a pest now, so what can I do now?'

There are many options available to help control pests, but if the options are ignored until there is only one option left (spray it) then obviously the farmer's options are extremely limited. For example, the options may include variety; time of planting; rotation; seed dressing; enhancement of biological control agents; and selection of pesticides in this and previous crops. The final choice of what can be sprayed is obviously extremely limited.

So the adviser needs to take into account not only the pest–predator situation but also the expectations of the farmer being advised. If, for example, the farmer wants a pesticide-based approach and wants to incur no risk to him or her, the advisor is less likely to recommend a wait-and-see approach. If, on the other hand, the farmer says that he or she is willing to wait and see how biological and cultural controls can perform, the adviser may be able to look for different things (such as the levels of parasitism and predation) and make assessments based on the level of biocontrol observed. There is a need for a collaboration between the farmer and the adviser for any change in practice to be made. With corporate farms, where there is a farm manager whose job includes pest management, there is an obvious incentive for this person to achieve immediate results and far less incentive to look for long-term sustainability, unless this is a long-term job or part of the farm plan or the overall aims.

It is up to the farmers who want to use an IPM approach to inform their advisers (including chemical resellers) as to what they expect and want in terms of pesticide advice. This is necessary so that the adviser knows that the advice required is not simply a decision about what to spray this week. Instead, the adviser will know that the information that he or she provides will be used in the context of a predator–prey model rather than a pesticide-based strategy aimed at eliminating all invertebrates.

A crisis in pest control

In many cases, the catalyst for making the change to something unknown (such as IPM) occurs when there is a crisis in pest control, because either the pesticides that have been relied on stop working (insecticide resistance) or the pesticide is no longer available (perhaps because its registration has been withdrawn or the product itself has been withdrawn from sale). While there are other factors that influence the decisions on whether or not to use a pesticide-based strategy (see Chapter 2), these two are the most common reasons for a sudden increase in adoption of IPM. Obviously, if the methods that are being used no longer work then something different must be done to control pests. In such circumstances farmers are more naturally responsive to a different approach. (Examples of different reasons for making changes are given in Chapter 9.) Equally, when a new pesticide arrives on the market that promises great and easy control, then the shift to using it as a mainstay of control rather than IPM is very high unless there has been a serious and bad experience of reliance on pesticides.

However, even in the absence of a crisis it is possible to achieve practice change and implement IPM. From our own experience it is entirely possible to have farmers understand and implement IPM without any particular crisis. Several examples of this are also provided in Chapter 9. *The key factors in the success of our work in adoption of IPM have been the collaborative and participatory approach to working with individual or small groups of farmers and providing expert, site-specific advice when required.*

However, finding an adviser who is willing and able to offer IPM advice (as outlined in this book) is not easy. Even when an adviser is available who offers IPM advice, it is necessary to assess whether or not the advice really is IPM or just pest management. So what can growers do? Why are there so few such advisers?

In the absence of a personal IPM adviser …

As described above, we believe that the best way to implement change is by farmers and advisers working together to obtain site-specific solutions. However, if this is not possible because of a lack of such advisers, farmers may still make great changes and advances. This is even more so if groups of farmers growing the same crop work together to find solutions. The first step is to understand the principles outlined in this book, in particular that:

1   Not all pesticides are safe to beneficial species
2   Pesticides safe to some species are toxic to others
3   A farm or crop is an agricultural ecosystem.

If these points are accepted, many advances are possible. With the current technology of the internet, email and digital photos available (with a constant promise of further developments) it is possible to find information and advice relating to the above points. There may not be absolute information regarding particular pest, crop, pesticide or location, but there is usually some information regarding the relative toxicity of pesticides to a range of species (for example, see <http://www.koppert.com> or <http://www.ipmtechnologies.com.au>). Using the available information, a farmer can begin to make changes to what has been the major decision involved in pest management: what should be sprayed? An option is to spray nothing. Information on what pesticides do to a range of beneficial species and not just to pests is the most important tool that most will need to begin to make changes to pest management.

Once the change in pesticide application has been trialled, the farmer may also be more interested in looking at what cultural controls can be utilised, along with the different spray regime. Again, information relating to cultural controls can be found, or potential methods trialled on farm without risking the entire crop.

There is an abundance of information available on insect pests, and in particular their lifecycles and identification. Some of this information can be easily applied, but it is not reasonable to expect every farmer to be an entomologist, so there will be limits to what can be achieved. This is where a specialist IPM adviser would be of great use. However, if the farmer is willing to trial control options and suggestions rather than simply follow recommendations provided by someone else, much can be done with a remote adviser. Importantly, not simply a web page with advice, but answers or suggestions to specific questions from a farmer about specific problems in a particular crop at a specific location and time.

Crop monitors and IPM advisers

To overcome the problem of correctly identifying not only all the life stages of pests but also all the life stages of different species of beneficial species, and distinguishing all of these from benign species that are neither pest nor biological control agents, there is the possibility of employing crop advisers. The crop advisers would then be required to have all of the skills required and the farmer need only to follow the advice. This is an appealing model, which follows what has been achieved in other areas of agricultural science (for example irrigation and crop nutrition). However, there is the possibility that the specialist advice regarding integrated pest management of invertebrate pests is not always as accurate as it could be.

For example, it is entirely understandable that a farmer obtaining advice on one topic or more from a respected scientific adviser (an agronomist or the person selling a product) would be interested in adding pest management into the range of topics for which advice is sought. This view is enhanced by scientists who like to include fungal and bacterial diseases, nematodes, weeds and all invertebrate pests under the heading 'pests'. So there is a range of advisers who offer different degrees of advice relating to control of invertebrate pests, but this may not be understood by the farmer asking for advice. Again, there is a requirement for the farmer to understand the type of advice to be requested.

The type of pest management advice offered by advisers can be put into several different categories:

1   The simplest advice is to use one pesticide over another without any assessment of pest levels and no monitoring of relevant species of beneficial populations. This could include prescriptive recommendations on pesticide use based on the calendar or crop stage.
2   The next step is for decisions to be based on pest numbers. Monitoring is carried out for pests but not for beneficial species. This is a step ahead of routine spraying, but obviously fails to utilise the full range of options.
3   When monitoring is done to assess the level of beneficials as well as pest levels, a more precise form of pest management can be offered. Greater skills are required of the adviser. If, however, only one pest and its natural enemies are considered, again this advice fails to utilise all options.
4   A greater degree of sophistication is required to consider all pests, all beneficials and all life stages, and to distinguish these from all other invertebrates. This requires a greater level of entomological skill, but for any given crop this is certainly achievable by those prepared to obtain the information. However, it may take time to correlate levels of pests and beneficials with levels of crop damage.

Crucial in each step described above is decision-making relating to the selection of pesticides. Decision-making becomes increasingly complicated with regard to pesticide application as you progress through each of the steps described above.

This book describes the options available in terms of pest management, and we have described how precise decision-making based on a range of pest levels, beneficials, cultural controls and selective pesticide applications can be achieved. So why are there not more businesses offering an IPM monitoring service? The answer lies in the relationship between farmers and advisers, and the reasons why such services exist in some industries in some locations and not in others.

Adviser–farmer relationship

This last point is extremely critical in the adoption of any strategy, but in particular the adoption of changed practice. What the farmer expects from an adviser, and also what the adviser thinks the farmer wants is what determines success or failure with any changed practice. If the farmer expects a pesticide recommendation, the adviser will probably produce such a recommendation. A recommendation to use no insecticide at all would be seen as high risk, even if there is a very good basis for assuming that biological control agents would achieve control. If there is a problem (for whatever reason) then the adviser may feel liable for giving the wrong advice, and the farmer may think the same, as he or she was expecting a different recommendation. What this means in practice is that the perceived 'safe' recommendation is to suggest the use of an insecticide, even if this may lead to later problems, such as secondary pests, insecticide resistance or pest flare (see Chapter 2).

Given that the main method of controlling pests is still by the use of pesticides, unless there is a collaboration between grower and adviser to trial something different, it is probable (at least from the adviser's point of view) that there will be an expectation to be advised to use a pesticide. If the advice is to not use a pesticide when pesticide applications are standard, if a pest problem occurs the adviser will feel vulnerable to litigation: hence the perpetuation of pesticide-based advice.

So in order to achieve change and to have advisers that are prepared to give IPM-type advice, the first step is for farmers to ask for that type of advice, at least on a trial basis, and for advisers to be prepared to have the necessary skills and decision-making to offer such advice. We hope that this book helps with the awareness of what is possible and helps promote that process of change.

Examples of failure

Our work with farmers does not always work out as we would hope. There are times when what we have to offer does not meet the grower's expectations and IPM is not adopted. However, there was an incorrect vision of what adoption of IPM could achieve within a specified time frame. For example, growers released *Persimilis* to control two-spotted mite but still sprayed broad-spectrum insecticides for insects such as aphids and western flower thrips. Too often when growers are using commercially available beneficial species, there is an incorrect assumption that they can use these beneficials like insecticides (see Chapter 1). So instead of basing IPM on biological control agents being well established in the crop, the grower believes that if a harsh insecticide is sprayed then a follow-up release of beneficials can immediately replace them. This is not so. Releases of beneficials are rarely *inundative* and more often *inoculative*.

Adoption of IPM relies on the farmer understanding (or at least accepting) that biological and cultural controls are the basis for effective control. The pesticides that may be recommended are only the support tools. The approach can fail badly when the grower sees only an alternative pesticide strategy being presented. This is particularly so when insecticides are recommended that are not as effective in the short term as others, but are recommended because they do not kill or affect key species of beneficials. The benefit of using such products is only seen when the beneficials are allowed to contribute and are not killed.

The reasons for beneficial species being killed or reduced in their effect are many and various, and most of these have been discussed in previous chapters. However, a recurring theme is that the full impact of pesticides on populations of beneficials is not generally appreciated. Even when the beneficials are not killed outright, there is often a reluctance to accept that pesticides can reduce the effectiveness of beneficial species by reducing fecundity (number of offspring) or affecting behaviour. This is most likely because these are very subtle effects when compared to pesticides that kill within seconds or minutes. A farmer used to seeing both pests and beneficials disappear within a day following the application of a pesticide is probably going to see the survival of adult ladybirds 24 hours after a spray as evidence that the spray is safe. This is obviously not true, but is understandable.

A much more subtle example of where an IPM strategy fails is where the same crop is grown, the same pesticide applications are used, no broad-spectrum insecticides are used, the same beneficial species are used and released at similar rates, yet one grower will fail while most succeed. Why is this so? It is obviously not an effect of disruptive insecticides which have an acute and immediately obvious effect (at least within one to two weeks). Sublethal effects are more likely to be the reason: the effects will not kill the individuals involved within a few days, but the effects can be seen in terms of reduced numbers of offspring or reduced predation or reduced lifespan.

Sublethal effects are much more difficult to assess and identify, but are just as real as more acute effects. The reasons for sublethal effects occurring are many, but are very difficult to quantify. The death of bees or colonies of bees worldwide currently falls into this category. Something is causing the death of bee colonies, but not necessarily immediate large-scale death of adult bees. However, the death of the colony is more important than the death of individual bees.

Examples of success

In Chapter 8 we provide a range of successful examples of changed pest management practices. We have chosen examples from the range of crop types in which we work to illustrate that success is not limited to any particular type of production. What is required is a desire to implement a strategy that is not based on pesticides and to have the necessary information on the biological, cultural and chemical options that are available. If not all information is available, on-farm trials are often the only way to find such details. The availability of an adviser in at least the initial stages of making change is extremely important, but this adviser does not necessarily have to be able to visit the farm regularly, or even at all.

Why success or failure?

The path to achieving successful adoption of IPM is very clear. It is well founded in the extension literature, and relies on farmers being involved in participatory trials (or demonstration trials) with direct access to IPM advisers. The IPM advisers must be able to give site-specific advice and not just generalities (such as information sheets or pest alerts). This approach has proven effective, even when there is no particular crisis in pest management, which is often cited as the only reason IPM is adopted. Instead, we suggest that the way to achieve IPM adoption is well proven, available and not restricted to any crop type or location (Horne *et al.* 2008).

However, this approach is not readily accepted by most of the agencies or organisations involved with pest management. Government agencies usually (certainly in Australia) have a policy of *not* providing site-specific advice. Extension staff have been replaced by Industry Development Officers charged with providing information on whatever particular grower groups require, but again, not site-specific advice. So advice from these agencies tends to be in the form of manuals, CDs and photo guides, which are all useful, but insufficient to actually make change happen in most situations. They increase awareness of possible change but not adoption of changes.

Growers can make changes on their own, find out information for themselves and implement it, but this is not common, and the number of growers doing this successfully for pest management is relatively small. Partly this is because the entomological skills and the information relating to pesticides are not easily available unless an IPM adviser is accessible. The adviser does not need to do all the work (such as monitoring) for the grower, but needs to be able to access help and support with training, identification and support at key times. Such support can now be provided much more easily than in previous years because of tools such as digital cameras and email.

Tools such as manuals on IPM or pest and beneficial identification charts can actually have the opposite effect to that which was intended by the authors, as some growers may think that if they need to identify everything in those manuals before starting, the task is too hard. However, once a new approach is started, and the grower seeks advice as required, then it is not as daunting as it seemed at first.

Once the decision to change has been made and the initial steps to use a biological and cultural control-based strategy have been taken, the task actually becomes easier. An adviser is not essential, but having one makes it much easier to find out about such matters as chemical safety and monitoring methods.

Two examples of pest control

Pesticide use in strawberry production (Victoria, Australia)

Strawberries are appealing to consumers because not only do they taste good but they look good too (plump, bright, red and shiny). So any pest or disease that upsets this look needs to be dealt with. It is worth noting that strawberry growers tell us that 10 years ago in Australia there was not the expectation of a perfect-looking strawberry. This is important, as the pursuit of the perfect-looking strawberry has overtaken other parameters such as taste. Quality assurance programs have factored in the cosmetic appearance, so these criteria must be met in order to supply. The main pest of strawberries for many growers is two-spotted mite, which has become resistant to many pesticides that are used against it. Fungal diseases can affect strawberries, so regular applications of preventative fungicides are registered for their control. Other pests that are of concern to strawberry farmers include aphids, some caterpillars, plague thrips and Rutherglen bug. In recent years another pest, western flower thrips, has become possibly the most serious pest for some growers and it also is resistant to the insecticides that can be used against it.

So how can the strawberry farmer defeat this range of pests? They include resistant pests as well as pesticides that when used against one pest kill the beneficials for another. There are no pesticides that will control all the pests, and any pesticides used must fit within the withholding period. Any residues must be below the maximum residue level (MRL). The simple answer is that control of pests in strawberry crops requires an approach that is not based on pesticides. Targeting specific pests with a selective insecticide (for example, aphids with Pirimor (pirimicarb)) is possible for most of the minor pests, but any insecticide sprayed will have an impact on populations of beneficial species. So minimising insecticide applications of any type is the aim in order to achieve better control of pests. This is a difficult approach to take when the experience for the farmer in recent years has been an increasing reliance on insecticides.

Control of brassica pests

Brassicas, which are grown around the world, include plants such as cabbages, cauliflowers, broccoli, pak-choi, canola and turnips. They cover the range of horticulture, broad-acre and forage pastures. However, the pests that occur are mostly the same in all of these (with some exceptions of establishment pests of broad-acre crops) and include diamondback moth, cabbage white butterfly and cabbage aphid. In the last few decades, diamondback moth has changed from being a minor pest to the most serious pest of brassicas because it has developed resistance to a vast range of insecticides, and is likely to continue to do so.

Although some have tried to talk of 'IPM for diamondback moth control', in fact what is needed is 'IPM for brassicas', of which diamondback moth is a key pest. Diamondback moth is a massive pest problem in many parts of the world, and many research projects have been conducted to try to develop effective control. In our experience, when it is considered as just another pest, with a range of natural enemies and cultural controls that can be used, with some pesticides used as support, control is fairly easy. We have seen this approach work in a range of crops across horticulture, broad-acre and fodder crops. It is important to use the biological and cultural control options first, then use the least disruptive pesticides as support tools. The basis of the pesticide support in our program has been BT (*Bacillus thuringiensis*)-based products such as Dipel and XenTari – not because of any greater kill rate compared to other products, but because of the minimal impact on beneficial species that help to control all pests. The use of these products will kill a large number of pest caterpillars and also leave intact the populations of predator and parasitoids that offer far greater control of the pests.

The combination of biological control and selective insecticides is an effective alliance that works to control a range of pests in brassica crops. Cultural controls, such as rotation, planting density, sequential planting and weed management, are also important. The farmer who grows brassicas in recent years would have experienced difficulties with control of diamondback moth in particular, and increased reliance on new insecticides. Changing to reliance on a product such as Dipel, which has been available for over 50 years, and beneficial insects that the farmer may never have seen is a difficult task to expect farmers to achieve alone.

Continuing difficulties

These two examples, controlling pests in strawberries and brassicas, highlight the fact that where there are insecticide-resistant pests there is interest in making a change in the approach to pest management, but it is difficult. The cultural control options in particular are often given low priority, leading to an overreliance on either biological or alternative chemical controls. There is a tendency, where commercially produced biological control agents are used for growers, to treat them as insecticides. However, in most cases the beneficials released are just an inoculation, and the population needs to establish and grow to achieve control.

Successful integration of the three available control options is the difficult part, and making assessments about the use and timing of each control measure is the key to whether or not the results will be successful. In our experience, often just waiting for one more week before intervening with even a selective insecticide can result in far greater levels of on-going biological control.

7

Changes in scientific assessment

In this book we have described the main issues that farmers have to deal with and the tools that they have to use. Other factors also influence the level of control that can be achieved, and changes in what is the accepted 'best practice' at any point in time. We generally accept that scientists using a sound scientific approach will provide advice that will give farmers the best means of controlling pests and this is the best way to progress any effort in pest management. However, we also need to accept that, while there are generally sound scientifically based principles underpinning agriculture, the development of knowledge about any given pest management strategy is not an absolute, immutable fact at any point in time, but instead is a continuum of progressive development. What was once accepted as the most scientifically sound, best practice can later be found to be not quite so effective. Over the centuries there have been many examples of how scientific thinking has developed and ideas have changed. At any point in time both the scientist and the farmer must make a decision based upon the best information that is available. Obviously the amount of scientific information that is available has increased over time, so decisions based on the best available information today may well be different from the decisions made on the best available information some years ago. Scientists of today would certainly argue that their research would hopefully result in improved knowledge that will result in changed practices for the benefit of farmers and others. This fact alone tells us that the scientific advice on most topics will lead to change.

Changes in scientific assessment and information loss

Over time, here can also be (and has been) a loss of information and skill. In the area of pest management this is possibly best illustrated by the sequence of interest in solutions offered. In the 1930s and 1940s there was interest in biological control and cultural control, and this was actively being researched. With the arrival of DDT and synthetic pesticides in the late 1940s and 1950s, the thrust of entomological research changed to investigating the possibilities of these new compounds. Even then, there was awareness by some entomologists of the impact that these new products could have on beneficial species. These warnings were lost among the more general desire to use and promote a new scientific-based (in this case chemical-based) approach. For example:

> In the event of any citrus growers switching over to DDT, adequate provision should be made by them … for the control of other insect pests, such as scale insects and mites. This is necessitated by the adverse effects which DDT might have on the natural enemies of some other citrus pests which, to some considerable extent are kept under control by these natural enemies (*Queensland Agricultural and Pastoral Handbook* **3**, 1951).

The potential of new products such as synthetic pyrethroids, which were very cheap in terms of dollars per litre of pesticide, advanced the role of pesticide-based strategies, a policy that remains in some areas to this day. Horticulture in Australia has now largely progressed past the stage of using such products, mostly because of insecticide-resistance problems, but broad-acre farming has largely not yet reached the same realisation.

Integrated Pest Management strategies have been the subject of much research effort for the last few decades, and are talked about by advisers to both horticultural and broad-acre farmers in Australia. There is no doubt that IPM strategies are further developed for horticultural crops, but the same approach has been successfully applied to broad-acre crops such as wheat, barley and canola in Australia. There are certainly examples of broad-acre farmers in Australia learning from the experience of horticulture (see Horne and Page 2008, and Chapter 8 of this book). However, these examples are the exception and not the rule.

Information that was valued by researchers in the 1930s regarding cultural controls was not considered particularly important by researchers focused on pesticides from the 1960s to the 1980s, but it was then considered more important again by researchers looking at the integration of biological and cultural controls during the last 20 years. This is a highly generalised version of history, but it is true that the focus shifted to pesticides in the 1950s, so those farmers, and the next generation of farmers, who grew up in this era have known pesticide-based strategies for their basic pest control strategy. Changes in the assessment of pesticides has led to the banning of many products that were once considered safe and were widely used. As new information becomes available (especially toxicological data), reassessment of products that are considered safe today may cause more products to be made unavailable.

There has been a change in the advice given by scientists, as the information available has changed. Those looking for new answers did not always seek out existing knowledge, and there has been a loss of information in pest management as one trend in pest management has superseded another. This will no doubt occur again in the future, as new information becomes available and leads scientists in new directions.

You may think that all information on the subject would be available on the Internet. However, unfortunately, the older information is not as available as more recent information, although this is changing. It is much easier to find the results of a trial 2 years ago (no matter what the publication) than something published or at least reported in the 1930s or 1940s. Reports that were not published in scientific journals but instead were in government reports, for example, are generally harder to find and retrieve. Obviously the scientists conducting studies in the 1930s did not have access to all the information that is available today, but modern research also needs to recognise the value in research done many years ago. It is sometimes surprising how much work has been done already, but the information is not readily accessible and the results are not used.

In this book there are three examples given of IPM being implemented successfully in broad-acre crops, including canola (Chapter 8). However, years after some farmers near Inverleigh in Victoria had successfully implemented IPM on their cereal and canola crops, there appeared an article in a respected scientific journal by scientists about the possibility of *developing* an IPM strategy in canola. Sometimes scientists are not as much on the cutting edge as the farmers and advisers who are collaborating in on-farm trials. Much can be done by on-farm collaborative trials to advance any aspect of farming, not just pest management. This was discussed in particular in Chapter 6. Farmers have a vast amount of experience, sometimes accrued over generations, and when they collaborate with scientists and participate in on-farm trials, great advances can be made. Often, scientists may simply be able to explain why something has occurred rather than develop a totally new approach. Great advances can be made when scientists listen to farmers and work with them rather than working in isolation, then offering a report.

Factors influencing successful control of pests

Here we look at some of these other issues that can have an impact on how effective pest management can be achieved. The first factor is the spray equipment and spraying methods used. As mentioned previously in this book, the basic decisions that a farmer has to make is (a) Do I spray an insecticide this week?; and (b) If so, with what?

If the suggestion is to spray product A, what else is needed by the farmer to achieve success? Once this decision has been made, many other factors will influence the degree of success that can be achieved. The most obvious of these is the spray equipment used to apply the selected insecticide. However, even with the best possible equipment there are other factors that can have a great impact on the effectiveness of the spray. Some of the factors include the time of day that sprays are applied, the temperature, the water volume used with the insecticide, the wetting agents used, the level of ultraviolet radiation and the incidence of rain or use of overhead or trickle irrigation. Also, these factors can have an impact on the beneficial species in the system: for example, the type of spray unit, the water volume used and the speed of the tractor.

What do we need to consider before the spray is applied? For example, soil preparation (or the conditions when the soil preparation was conducted) can have a massive influence on the degree of control achieved by pests such as potato tuber moth in potato crops or slugs in canola crops (see pages 44–45). The impact of the previous crop can also be highly important. If there is rotation through different crops, the relative pest pressure from some pests (and diseases) can be greatly reduced. If the pest or disease does not have a suitable host for a considerable period of time, there will be less carry-over of both pest and disease problems. So control of pests and diseases that are associated with particular crop types (such as brassicas) can often be made much easier, and the reliance on chemicals made much less, if rotation through different crop types is implemented.

Future strategies

Because a range of biological, cultural and chemical options will be offered in the future, it is important to adopt a compatible set of control options, rather than provide a rigid set of rules. We simply present the current (and past) situation, in the hope that readers will be able to see how a pest management strategy can be put together and implemented now and in the future using the tools available at the time.

Control of potato–tomato psyllid

Tomato–potato psyllid is a tiny insect that looks like a very small cicada. It lays eggs on short stalks, and the scale-like juvenile stages suck the sap of host plants. It is such a serious pest that it has been reported to cause up to 80% crop losses. When it arrived in New Zealand, the first reaction was to try to eliminate it with insecticide applications. This failed, but this course of action is still likely to be followed by Australian authorities who will deal with it (and will almost certainly fail) in exactly the same way whenever it arrives.

This pest, currently in New Zealand, is devastating potato crops and also hydroponic capsicums and tomatoes. The accepted control strategy in field potato crops has been to treat it very heavily with insecticide sprays. One currently available best-practice strategy from advisory bodies is weekly insecticide applications along the lines described below:

> Actara as a seed dressing, then sprays, each week as follows: Nitofol/Nitofol/ Karate/Karate/Vertimec/Vertimec/Nitofol/Nitofol/Karate/Karate. (Nitofol is a broad-spectrum organophosphate insecticide, Karate is a broad-spectrum synthetic pyrethroid insecticide and Vertimec is an insecticide that is not so residual but is also broad-spectrum. All will kill the key beneficial species in potato and other solanaceous crops.)

We believe that a much better and sustainable control strategy is available and applicable. However, to trial it there is a requirement for farmers to implement IPM or at least to believe in the approach. This was impossible initially, as all farmers sprayed insecticide for psyllids. However, in 2010–2011, a few advisers and farmers were prepared to work together, with on-farm trials in commercial crops, to try something different. Rather than sit back and wait for entomologists to conduct their research and deliver them the answer, farmers are actively taking part in large-scale trials to help find an answer. The results of this new approach, which involves using resident biological control agents, new cultural controls and only two or three selective insecticides, has so far been positive.

We expect (in May 2011) that an IPM approach based on using a compatible set of control measures that are based on biological and cultural controls, and supported by only strategic use of highly selective insecticides, will be the basis of the control of this pest in the future. It is not yet even a pest in Australia, but we offer this as the test for future readers of our approach in this book.

8

Examples of changing pest management: specific crops

Changing to IPM

Our work with farmers usually involves assisting them to make the change to using IPM. We draw on our experiences to offer a range of examples from different crop types, and also from other advisers and pesticide retailers and producers. We also outline where the change did not work successfully, and we suggest why that occurred. The examples are from our own experience, so the crop types reflect the places where we work and the pests to be dealt with in those crops. However, there are enough similarities to suggest that the approach to making change is what they have in common and that the approach will work anywhere with the right support. The emphasis in these examples is the process of making change successfully rather than depending only on the IPM strategy. The change involved may well have been made from any chemical-based strategy to another instead of changing to IPM.

Certainly there are very many examples of IPM working well, or we would not be writing this book. IPM can work, but it can also fail terribly if the grower has mistaken expectations of what is required. The worst examples are when a mixture of old pesticide applications and introductions of beneficial insects and mites are made; that is, when neither one approach nor the other is selected. Basically, a grower should use IPM or rely on sprays (that is, not use IPM), but an inappropriate mix is doomed. If compatible sprays can be incorporated there may be a better outcome, but the number and type of sprays used are relevant. One spray at the label rate may be compatible, but multiple sprays at or above the label rate can be highly disruptive to an IPM strategy.

Some of the examples given here are from articles already published in magazines or other literature for farmers. We begin with an example of where changing to an IPM approach did not work out well for the farmers, then give some examples where change was implemented successfully.

Where a change to IPM did not work

Ornamental flowers: Sietze and Peter Baas, Monbulk, Victoria

Sietze Baas and his father Peter are commercial flower growers using about one hectare of plastic polytunnels in Monbulk in Victoria. They specialise in growing varieties such as Bells of Ireland (which accounts for about 80% of their production) as well as Statice and Canterbury Bells. They have grown these varieties for over 12 years.

The main method that they had used for many years was insecticides. Similarly, control of disease problems was mainly through application of fungicides (in a regular program, rotating through different chemical groups).

The main reason for wanting to change to a different method of pest control was to benefit the local environment and the farmers' health status by using less pesticide. They had heard of IPM, and started to research it for themselves. The first step was to find out the chemicals (insecticides, miticides and fungicides) that could be used with the beneficial insects and mites that they needed.

They began to release *Persimilis*, *Hypoaspis* and *Cucumeris* to control two-spotted mites and western flower thrips. They also found that they were able to use naturally occurring *Aphidius* wasps and brown lacewings to control aphids. It was necessary to change the fungicide program, and in particular to stop the regular use of Mancozeb and Bavistan.

Control of mites, aphids, caterpillars and thrips all worked well using IPM. Sietze noted that instead of having to spray for aphids about every 3 weeks in the warmer months, he did not have to spray for aphids at all. However, in 2010 the weather was unusually humid, which caused problems with a fungal disease called *Cercospora*. The Bells of Ireland originate in Syria and Turkey, where humidity is low, and conditions in Victoria in 2010 were very different. Disease problems began to cause severe losses. The Baases resorted to the fungicides that they knew would control the problem. Unfortunately these fungicides killed the beneficial insects that they had used within an IPM program, so they then needed to resume spraying for insect pests.

Sietze said that he was very happy with the control of insect pests, and may try using IPM again, but only if there was a better way of controlling key fungal diseases.

This example is not common, as usually control of fungal diseases is easier than control of the invertebrate pests. However, the same weather conditions also caused problems for leek growers, who would not normally have to spray for a fungal disease called purple blotch.

'We tried IPM and it did not work'

We hear this statement quite often from a range of growers of very different crops. In most cases the growers honestly did try to implement IPM but did not succeed. Why did it not succeed? The most common reason is that they did not actually implement IPM, although they thought they were doing so. One of the most common mistakes that is made is the thought that buying and releasing beneficial species is IPM. It may or may not be necessary to do this, but integrating the use of beneficial species with other control measures is essential.

Lack of knowledge, or not being able to access information about pesticide effects on beneficials, is also a very common cause of IPM not being successful. On one hand growers may be releasing beneficial species that they have bought from commercial insectaries, but on the other hand they are killing them or setting back populations with pesticide applications. It has also happened that when relying on releases of commercially produced beneficials that they attempt to switch over too much area at once to IPM. This is because the number of beneficials is initially low and the pest can flare before the population of beneficials builds up enough to give control. Getting control in one area at a time can be a better option in such cases.

There is a range of pesticides that are not broad-spectrum, and can have a place within IPM strategies. However, growers can make the mistake of simply rotating their chemical applications through the list of 'safe' products (see Chapter 2). This is to confuse an Insecticide Resistance Management (IRM) strategy with IPM.

Successful examples of change to IPM

Here we describe how farmers and others have changed successfully from using a conventional, pesticide-based approach to dealing with pests to one using IPM. We use these examples because they are (mostly) examples that we know and have helped to make happen. However, they are certainly not exclusively from our own experience. The first example is one that is outside our experience, but one that proves the point: IPM is an approach that can work in any crop, anywhere. The example of control of potato pests, as achieved by Wayne Tymensen, is one of the most impressive and durable (16 years so far) and should give hope to those wanting to obtain sustainable control of potato pests worldwide.

Avocados: Bonnie and Tony Walker, Alstonville, NSW

Bonnie and Tony Walker own a 4-hectare avocado farm near Alstonville in New South Wales that produces an average of 70 tonnes of avocados per year. Before using IPM, control of pests and diseases was achieved using a calendar-based spray program. Chemical insecticides were used to target pests, and copper-based fungicides were used for disease control.

However, Bonnie became concerned about possible effects of copper on soil health and the insecticide effects on non-target insects and mites. Courses on sustainable agriculture prompted interest in changing control measures, so Bonnie began to try an IPM approach, starting with just a small part of the farm and progressively increasing the area as confidence in the approach grew. It took Bonnie and Tony 3 years to move to an IPM approach for insect control over the entire farm, but longer (8 years) to eliminate the use of copper sprays.

The first step was to understand not only the lifecycles of the pests but also those of the beneficial species involved. Changes that were made included using insecticides that were selective. The first used was Dipel, based on *Bacillus thuringiensis* or B.t. Knowing how and when to apply B.t. was an important step in adopting IPM. Another significant change was the use of commercially reared *Trichogramma* wasps. The use of these wasps for biocontrol of caterpillars helped to change from a pesticide-based approach, but they are now no longer needed in the avocados. However, they are still used in the macadamias that Bonnie and Tony grow. It helped greatly to have the advice of a consultant who was familiar with organic methods of production.

The advantages of having made the change to using IPM now include healthy soils, and the fact that chemical pesticides are only used in hot spots, as and when required. In addition, control of pests is now better than when using a pesticide-based approach, as control measures are now targeted more precisely. Decisions are based on monitoring as well as historical knowledge of the pests and beneficials.

Broccoli: John Fabbian, Werribee South, Victoria

John Fabbian grows broccoli, fennel and lettuce in Werribee South in Victoria. On his farm, *each year* he plants in total about 32 hectares of broccoli, 16 hectares of lettuce and 20 hectares of fennel. Until 2000, John controlled insect pests with a regular spray program using a range of broad-spectrum insecticides. Spraying needed to fit in with the weather or watering, and monitoring for pests was not always carried out.

In 2000 John decided to trial an IPM approach on his lettuce crops as an alternative to a pesticide-based program, simply because the pesticides he was

Figure 8.1   John Fabbian in his crop of broccoli

using at the time were not working, mainly because of insecticide resistance in *Heliothis* caterpillars. The initial change to try something different was seen by John and some of his neighbours as very risky. 'There was no one in Werribee using IPM, so we could not really see for ourselves how it would work.' So John was one of the first growers in Werribee South to change to using IPM.

The initial results in the lettuce were good, which gave John (and others) the confidence to start using IPM in his broccoli crops as well. Confidence grew as each successive planting came through with good results, and John grew to trust the approach being taken. One part of the change was looking for eggs, and timing any sprays at small caterpillars, as well as using 'good bugs' to help control the pests (and in particular *Plutella*). For many years now it has been the only way that John will deal with pests in his broccoli.

John has his crops monitored weekly for about 7 months of the year, looking for both pests and beneficials, and decisions on any action are made based on the results of that monitoring. Having independent advice and a good relationship with people he trusts is essential to John, and it is one less job that he needs to do. The fact that his advisers (Paul Horne and Jessica Page) do not sell chemicals is also important, as is the access to information about new products that they provide.

The results have been better control of pests, with less pesticide being applied. John notes that since changing to using IPM he may now only spray the broccoli crops for aphids once per year, but in the past it was required almost weekly during the warmer months. In addition, 'not having to handle S7 chemicals is a real bonus'.

Broad-acre cropping: Rowan Peel, Inverleigh, Victoria

When we first talked to broad-acre farmers in southern Australia about IPM, many thought that it could only be applied to horticulture. However, a group of farmers in Victoria have shown that exactly the same principles can be applied to broad-acre crops, such as wheat, barley, canola and field peas. Instead of applying cheap insecticides, control is based on naturally occurring beneficial species and cultural control methods. If chemicals are used, selective products or seed-dressings are preferred.

Rowan Peel, who runs a cropping and grazing property near Inverleigh, was the first in the group mentioned that began to use IPM in broad-acre in Victoria. The main farm is 1350 hectares, plus 400 hectares run as a share farm. The crops are a rotation of wheat, barley, canola and lucerne. Before using IPM, Rowan had a fairly standard, calendar-based pesticide strategy (using broad-spectrum insecticides). That consisted of applying insecticide with the herbicides before planting and just after sowing, then spraying for aphids at predetermined times, and possibly for grubs such as *Heliothis* late in the season. Baiting for slugs was also standard in canola crops.

His approach to pests and pesticide use has now changed totally. 'We now only use insecticides if absolutely necessary, and then we try to use selective products. We began by trialling IPM on three paddocks in 2003, but quickly decided that this was the way to go, and in 2004 decided to apply IPM on the whole farm. In the last 5 years we have not used an insecticide through the boom spray!'

Rowan sees several advantages in using IPM:

> The main advantages to us are being better off financially, not having to handle so much pesticide, it is better for the environment, and also we know exactly what pest we are dealing with and so are getting better control. Farmers like to see other farmers in their area having success with any new technique, and so on-farm demonstrations are a good way for growers and agronomists to become confident in IPM. It also needs skilled people to help get it started. The decision-making became much easier when we began to understand about beneficials in our crops. That meant having a really good look in our crops and being aware of what was going on at that level on our farm.

Broad-acre cropping: Ian Waller, 'Mooramong', Skipton, Victoria

(The article below was written by Fleur Muller for the publication *Farming Ahead*, published by the Kondinin Group (based in Bentley, Western Australia) in 2010. Used with permission.)

IPM reduces reliance on insecticides

Farm information

Farmer: Ian Waller

Location: Skipton, south-west Victoria

Property: size 1500 hectares.

Enterprises: Cropping, (wheat, barley, canola, peas); 5000 fine-superfine Merinos; lucerne.

Annual rainfall: 525 mm.

Insecticide costs have been drastically reduced on one Victorian mixed enterprise property since adopting integrated pest management (IPM).

Ian Waller manages 'Mooramong', Skipton, for the National Trust, and has been using IPM strategies for about six years. For him the move is paying off with clear economic and management benefits.

The annual cost of the IPM program is now about one-fifth of the annual cost of applying insecticides. Crop damage has been reduced, the risk of developing insecticide resistance is low and beneficial insect populations have increased and are more in balance with pest populations. A 200-hectare nature reserve, situated in the middle of the property, has also benefited from the new IPM strategy and has become a valuable resource in the IPM approach.

'We were traditionally frightened if we didn't use insecticides we would be likely to strike a problem,' Ian said.

'We would always carry out two sprays, usually blanket sprays of crops annually, including long-acting broad-spectrum residual insecticides in canola to control pests such as redlegged earth mites, aphids, slugs, army worms and Heliothis.'

Ian found the approach was not working. Pest numbers actually rebuilt more quickly and their beneficial predators were being killed off and their populations took much longer to rebuild.

'We were only getting a short-term benefit from spraying and were looking for a better way to control pests.'

Getting started

'We began with three paddocks, with three different crops which were monitored by Neil Hives, as part of a trial run by IPM Technologies. By enlisting the services of an entomologist, we quickly gained the confidence to no longer depend on insecticides.'

This initial experience convinced Ian to apply IPM principles across the whole property, and so far there have been no disasters or economic yield losses. Redlegged earth mites have not been a problem in canola apart from along the paddock edge, up against pasture. Slugs have been a problem on the cracking clay soils in some years, but putting tiles down before sowing has helped Ian monitor hatchings. This has also enabled him to predict if baiting is necessary before a problem arises.

Insecticides are now only used if absolutely necessary, and to date border sprays, baits and seed dressings have been the main chemical control methods applied. Selective insecticides that, as much as possible, target the pest while leaving beneficial populations to attack and suppress the pest, are the first choice. While they are generally more expensive per hectare than broad-spectrum insecticides they are more cost effective in the overall pest control system.

Identify and observe

Ian says careful observation and regular monitoring during the season is the most crucial part of a successful IPM program and are best carried out by an expert. Ian employs a private entomologist to monitor paddocks during the growing season, believing correct identification of pests and beneficial species are the most important aspect of any successful IPM strategy.

'Simply reducing insecticide use won't work if you don't monitor the level of pests and beneficials. Low levels of pests can be tolerated without economic damage.'

Starting shortly before sowing, monitoring continues on a weekly basis for about two months after sowing, then every second week through the growing season until shortly before harvest. Decisions are made based on the relative number of each pest. There may be enough beneficial insects to keep pests below damage-causing thresholds.

'It is surprising to observe the large diversity of beneficial predators available in the paddock and the swarms that follow the pests, such as aphids, that can control them given the chance. You really need to provide the beneficials the time to do their work.'

Changing diversity

Since adopting IPM some insects not previously identified as crop pests have appeared, requiring targeted control measures. Ian believes pests, such as earwigs, which were first identified in canola during the first year of IPM, were possibly destroyed by past spraying activities. Insecticide-treated grain has proved successful against earwigs when populations have surged beyond threshold targets.

A new pest mite is currently causing major damage in Ian's district and is proving very difficult to control with any insecticide. But while it has been identified in low levels on the property Ian says the mite is yet to cause any damage.

'I can only surmise at this point that the large number of beneficials present is keeping them at bay. Successful IPM takes time and persistence.'

Ian's IPM strategy

- Correct identification: an experienced entomologist is paramount
- Treat canola seed with insecticide such as Cosmos
- Identify potential problem paddocks early and monitor with tiles
- Make weekly paddock assessments through the early stages, to record pest and beneficial populations
- Monitor aphid flights arriving during spring, and assess beneficial predator numbers versus pest numbers
- Check armyworm and heliothis numbers during late spring together and consider the control efforts of the parasitic wasp. Consider safe control options, such as Dipel, if control is needed.

Broad-acre cropping: Colin Hurst, arable cropping farmer, Canterbury, NZ

Colin Hurst runs a 700-hectare family farm near Waimate in the southern Canterbury region of New Zealand. The main crops that he grows are wheat, grass seed and brassica seed, and he also grazes sheep. Cropping accounts for 60% of the farm area each year.

Before 2006 Colin was operating a conventional system using synthetic pyrethroid sprays (Karate) as the main defence against aphids, and the barley yellow dwarf virus that they can vector. However, in 2006 Colin had an introduction to the concept of IPM via a project initiated by the Foundation for Arable Research. The initial discussion that aroused Colin's interest was that the sprays applied for aphid control could be causing disruption of the control of other pests, and in particular slugs. In wet years slugs were a significant problem and were expensive to deal with (using baits).

The idea of an IPM approach sounded interesting, especially as it offered an alternative to overreliance on a single insecticide and the development of resistance, but it was also very different to the mainstream approach being used at the time in New Zealand. The idea of seeing pests increase in number but not applying an insecticide spray was one of the biggest changes to be made, and one of the biggest concerns in the early stages. So Colin trialled the approach on half a paddock. This allowed him to see the results of each method and compare the results not only in terms of cost of pesticides (including baits) but also in terms of yield.

The first year's trial was very encouraging: no pest issues, reduced pesticide use and an increased awareness of beneficial species. Colin had back-up with insect monitoring and decision-making during this time from Plant and Food Research (a government agency in New Zealand) and IPM Technologies P/L in Australia as a part of the project. The realisation that there were many beneficial species present in his crops was something that Colin had not utilised before, so he decided that he should investigate further.

Therefore, in the next 2 years of the 3-year project, Colin progressively adopted an IPM approach, as he felt more comfortable with the new IPM strategy and decision-making based on monitoring. The monitoring that he uses consists of direct searching for a range of predators and parasitoids, sticky traps to assess what is flying at any time and aphid flight information provided by the Foundation for Arable Research (with Plant and Food Research). Using IPM has meant a much greater use of monitoring than in the past, so any action is based largely on observations on the farm rather than predetermined sprays or district-wide information. Colin now uses seed dressings of synthetic insecticides rather than sprays as a part of the strategy to use minimal insecticides. Instead he relies on cultural and biological controls.

The change has been dramatic. In 2010 (4 years after implementing change) the only slug problem requiring treatment was on a border with a neighbour's field, where there was invasion from outside the farm. The only control measure required in this case was a border application of slug bait. Colin now believes that carabid beetles provide a significant level of control of slugs, and is keen not to disrupt this control with sprays targeting aphids.

After 3 years of trials, Colin now implements an IPM strategy over the entire farm, and only uses selective insecticides to support the biological and cultural controls as necessary. Although initially daunting, the change in practice has proved worthwhile.

Capsicums: Henderson Hydroponics, Tasmania
(This is the basis of an article published by *Good Fruit and Vegetables* magazine in 2009.)

Rob Henderson (and his family), who grow hydroponic capsicums near Devonport, Tasmania, experienced a major problem during the 2007–08 season with tomato spotted wilt virus (TSWV). This virus affects the plant and the fruit, causing affected fruit to be unsalable. The only treatment of infected plants is their removal and disposal, which resulted in approximately 75% of TSWV-susceptible cultivars being removed before the end of the season, which resulted in considerable financial pain. The problem virus is spread by several species of thrips, including western flower thrips (WFT), which had not previous been present at Henderson Hydroponics. WFT are resistant to many insecticides, which was the problem in this case. The thrips were surviving the insecticides that were used, so were literally out of control. Rob needed to do something different to manage these pests, and insecticides did not look like the answer.

In 2008, before his latest crop was planted, he met with Dr Paul Horne and Jessica Page of IPM Technologies to discuss implementing an IPM approach. Paul and Jessica were in Tasmania to help develop IPM in a range of vegetable crops, and were introduced to Rob by an agronomist (Peta Davies from Roberts Ltd) who saw that this may be the answer to their problem.

A range of predators were introduced throughout the season to control fungus gnats, WFT, aphids and two-spotted mite. It also meant that the broad-spectrum insecticides that he had used in the past could no longer be used, and extreme care had to be taken to ensure that the beneficial species were not disrupted by attempts to control other pests.

Rob admits, 'We were sceptical at first about IPM, but now we are converts.' He said, 'The western flower thrips were present in the latest crop but the predators, in time, controlled them and total damage was reduced to below 2% infection, down from 75% the previous year.' Two-spotted mite was becoming an increasing problem in previous seasons. However, excellent control was achieved with the release of *Persimilis*, which displayed a ravenous appetite for two-spotted mite.

There were some very nervous moments early on in the season when WFT were obviously present and before the predatory mites (known as *Cucumeris*) had taken control of the pests. However, the results later in the season speak for themselves, and the next season's expanded crop will again be grown using IPM, but with less nervousness now that Rob knows what to expect.

Rob will be trialling a new thrips predator called *Orius*, which is a new possibility for WFT control. It is being produced in Western Australia.

The project that allowed this to happen is funded by Horticulture Australia and the AusVeg levy.

Rob estimates in his first year of IPM he would have spent approximately treble that which he would have normally spent on insecticide. However he is hopeful that next season, having had a season of IPM experience behind him, this cost may reduce. However, as a qualified agricultural economist Rob considers the expenditure on IPM in his greenhouses to be an extremely sound and profitable investment both financially and environmentally.

Henderson Hydroponics staff enjoy working in an environment free of insecticides, and have noted the dramatic increase in natural predators, particularly frogs and ladybirds, seen this season in the greenhouses.

Henderson Hydroponics customers have also been pleased to purchase quality fruit grown without the use of insecticides.

What has Henderson Hydroponics learnt from one season of IPM?

- IPM does work.
- Constant crop monitoring is very important.
- Don't be afraid of seeing small numbers of pests, leave the insecticide locked in the chemical store and only think about using it as a last resort.
- Predators take time to multiply. If pest numbers are increasing, purchase and release more predators rather than waiting for predators to breed up.
- Good advice is readily available. Make use of it. Henderson Hydroponics are extremely grateful for the advice provided by Paul and Jessica during their visits and by them being available to answer questions on the telephone and email at short notice.

Celery: J. and J.M. Schreurs, Clyde, Victoria

J. and J.M. Schreurs are the largest producers of celery in Australia, planting about 5 hectares of celery per week, and harvesting around 20 000 boxes per week.

Production of celery on this farm for over a decade has relied on IPM. This is because a previous reliance on pesticides failed and IPM has been proven. Control of caterpillars and aphids that vector viruses were initially the key pests to be dealt with, but chewing pests, such as vegetable weevil, lightbrown apple moth and cutworm, are also equally important at certain times. All of these pests needed to be dealt with within an IPM strategy and it has been achieved.

Theo Schreurs, his brother Tom, and son Adam are responsible for controlling the key pests and diseases that the crop may face. In the past the approach was to apply pesticides to control both pests and diseases. They tried something different because the pesticide-based approach had failed at the time. 'We were spraying more and more insecticides and there was still damage and loss. We had to look at another approach,' Theo said. Theo and Adam had talked to their potato-growing neighbour, Wayne Tymensen (see separate story), who had already experienced the change from relying on insecticides to using an IPM approach, and he encouraged them to contact Paul Horne about what could be done.

Initially sceptical, Tom, Theo and Adam reached the point where they had to acknowledge that what they were doing to control pests was not working, and that they had nothing to lose by trying what their neighbour recommended. Because the situation was so serious, they decided to change the entire farm, not just a paddock.

The new (IPM) approach worked well on the target pests, although learning about what were the real pests and how and when to deal with them took a while, and there were losses at each step as each separate pest was identified and methods to deal with it were developed. Pests that had never before been experienced (because they had been killed by broad-spectrum insecticides targeting primary pests) suddenly needed to be dealt with, because the selective methods of dealing with the primary pests did not control the secondary pests. Each pest has been dealt with, but it was only the collaboration between the Schreurs and IPM Technologies that has allowed the full range of pests to be identified and dealt with.

The use of an IPM approach on all of their farms has meant significant changes in how and when insecticides are applied. Insecticides are now selected and used only as required, which has meant that far fewer insecticides are used than before an IPM approach was adopted. Now decisions are made about any insecticide application, and what is selected, if any, on the basis of monitoring. The monitoring and decision-making now includes the level of biological control, the impact of cultural controls and the available chemical options.

Now there is an awareness that beneficial species can achieve better control and that pesticide applications need to be carefully targeted. In particular, Theo, Tom and Adam are aware that pesticides applied for one pest can interfere with control of another pest by disrupting the populations of beneficial species. That includes pests, such as western flower thrips, that are not a problem in their crops because their control methods do not allow them to become established. Theo commented, 'There has not been a flare-up of aphids since we have begun using IPM.' Aphids still need to be controlled, but they have never been out of control since adopting an IPM approach.

Theo explains that the main problem that they have had with using an IPM approach has nothing to do with pests or damage. Instead it is the fact that, currently, consumers and wholesale buyers can object to the presence of any insect: good, bad or benign. That is, if an insect is found in a product on sale in a supermarket, it does not matter whether or not it is a pest causing damage, a beneficial species that helps control pests, or any other insect that may have just been passing through.

The benefits of using IPM now include better control of pests and reduced reliance on insecticides. Theo is convinced that the change that they made was worthwhile, and the control of pests of celery will from now on always be based on IPM.

Citrus fruits: Ingerson Family Citrus, South Australia

(Based on information supplied by James Altmann, Biological Services, Loxton, South Australia.)

The Ingerson family from Berri and Bookpurnong in South Australia has entered a fourth generation of citrus production, having grown, packed and marketed their own produce since 1931. Today they run 110 hectares of citrus, mostly navel oranges, but including mandarins, ruby red grapefruit, lemons, limes and valencia oranges.

Principal David Ingerson says that all generations have had strong interests in sustainable production right from their early beginnings, with environmental awareness being at the forefront of their thinking.

Biological services supply *Aphytis melinus* wasp parasites, which are released regularly to aid in the control of red scale, and Fruit Doctors consultants have monitored pests, disease and beneficial arthropod levels since 1987.

When the Australasian Biological Control association floated the idea of an IPM accreditation in the late 1990s, David was immediately interested in utilising the logo to identify their fruit to the public as a 'clean–green' product predominantly free from pesticides.

To promote other beneficial species in the orchards, a range of broadleaf and grassy plants are allowed to grow in the mid rows between the trees. This supplies nectar sources for wasps, pollen for predatory mites and alternate hosts for generalist predators such as lacewings and ladybirds. Ingerson citrus has also supported research projects for the establishment of new key beneficial species over the years. David helped to instigate the introduction of parasites for *Citrophilus* mealybug, and has regularly offered the orchard for trials for release of wasps, and leafminer to help control soft scales and thrips. In the harvest and packing shed the Ingersons will also tolerate some blemishes on fruit from pests, rather than trying to grow perfectly clean fruit with pesticides.

Since 1991, the ABC IPM logo has been proudly displayed on all of the 'Ingy's' citrus packaging as part of their marketing strategy, and to help promote the concept of producing safe, healthy fruit without the need for regular toxic pesticide applications.

Leeks: Peter Schreurs and Sons, Devon Meadows, Victoria

Peter Schreurs and Sons grow a range of vegetable crops, including leeks (the main crop), lettuce, endive, kohl-rabi, wombok, radicchio and parsnips, on their 160-hectare farm in Devon Meadows near Cranbourne, Victoria.

Darren Schreurs first encountered IPM when he was trying to deal with mites and thrips in his leek crops. (For the full story see <http://www.leeks.com.au> or Horne and Page 2008). He was trying to use insecticides and miticides to eliminate thrips and mites, but it was not working, so he adopted an IPM approach. The IPM approach was highly successful, and a decade later IPM remains the mainstay of pest control on all crops on the farm, not just leeks.

However, so far as the main crop on the farm goes, since adopting an IPM approach to dealing with pests in his leek crops Darren has not applied a single insecticide on his leek crops in the last 10 years. This compares to two insecticide sprays each week just before adopting IPM. It is worth pointing out here that the leeks grown on this farm include exports to Japan, and the lack of insecticide applications is an advantage when dealing with buyers, but the main factor is the high quality of the leeks.

However, on a local level, one of the biggest problems that the farm has with using IPM is that there are still live insects in the crop that are not damaging pests, and may be beneficial, but wholesale buyers reject the produce if any live insects are present. So there is pressure from supermarket buyers for an insecticide application just before harvest, which is something that the Schreurs do not want to do. They find it very frustrating that the crop can be grown with minimal use of insecticides and be rejected because of a non-pest insect being present.

Lettuce: Fred and Stan Velisha, Werribee South, Victoria

Fred and Stan Velisha run a business (Velisha Brothers) in Werribee South where they grow iceberg lettuce, cauliflowers and celery. The iceberg lettuce production is unique at the time of writing (2011), as they are the only farmers in Werribee South, as far as we are aware, who arc using an IPM approach to lettuce, which to us means in particular choosing not to use a seedling drench of *Confidor* (imidacloprid).

Each year Velisha Brothers produce around 120 000 cartons of iceberg lettuce (with each carton containing 12 lettuces). Before using an IPM strategy to control pests in lettuce, Fred and Stan used a regular spray program of broad-spectrum insecticides in the field. When lettuce aphid arrived in Australia, the accepted method to deal with it was with a seedling drench of Confidor (imidacloprid), but at a rate that was around 18 times the rate used as a spray in the field. This rate was a problem for IPM, as it killed beneficial species that could control the aphids by secondary poisoning.

The insecticide drench was (and still is) an impediment to adoption of IPM, as it kills the species of beneficial insects that would otherwise give control. However, it was also a problem for Stan in the nursery. He applied the recommended dose to seedlings in his nursery for a couple of plantings, but then decided that health and safety issues were paramount, and that the dose of imidacloprid required to achieve control of lettuce aphid was not something that he wanted to do or offer to his clients. The decision to not apply the imidacloprid seedling drench has cost Stan customers, but it is a decision with which he is very comfortable, as he needs to balance safety and efficacy.

The benefits to Fred and Stan include the better control of pests in the paddock, with less chemical use, but also the more precise use of even selective insecticides. IPM is now an important part of their production methods, and the only disadvantage to them is that pesticides targeting other pests (caterpillars) are best applied late in the day to avoid peak times of UV.

Macadamia nuts (1): Doug Rowley, Rosebank, New South Wales

Doug Rowley has owned Brushbox Farm at Rosebank in the Northern Rivers area of New South Wales since March 1994. Doug comes from a non-farming background, and was an absentee owner until late 2003. The farm area is 50 hectares, of which 17 hectares is devoted to macadamias, the balance being a big scrub remnant and revegetation by native and non-native species. The 4200 macadamias vary in age and variety; 35% are 22 years or older and are on a 7 metre by 7 metre spacing, 40% are 15 years old on an 8 metre by 4 metre spacing, and 25% will be harvested for the first time in 2011 and are on a 10 metre by 4 metre spacing. The orchard has been divided into six blocks, and production is recorded in tonnes per hectare (and kilograms per tree) over each of the blocks. In the 2010 harvest, the best block produced 4.2 tonnes to the hectare and the worst 2.9, with the average being 3.8.

When Doug purchased the farm he decided IPM practices were a totally appropriate way to manage pests while minimising collateral damage caused by inappropriate spraying of insecticides. Therefore, Doug never had the experience of using a non-IPM strategy. However, the example is included here as one that exemplifies that IPM works in the absence of a crisis in pest management. For the approach to work, Doug relies totally on pest scouts to monitor activity levels and to recommend appropriate courses of action. The scouts, subject to weather, check the farm every 3 to 4 weeks from the commencement of flowering, usually from July to December. Spraying, when necessary, is done with a double-sided air blast sprayer fitted with gap-sensing sonars that automatically switch off the spray in the gaps between trees.

Doug deals with three major pests: lace bug, fruit-spotting bug (FSB) and macadamia nut borer (MNB). Invariably these pests present in small areas, and effective control is usually achieved by spraying these 'hot spots'. Lace bug, which can completely destroy a crop if left unchecked, is easily controlled using this technique. FSB is more difficult to control. As soon as it is detected the hot spots are sprayed, but Doug is dissatisfied with the level of control achieved. As a result he is currently looking at cultural control options based on better monitoring, planting murraya hedges and a different spray regime that is still solidly based on IPM principles. MNB, once controlled by spraying, is now totally controlled by *Trichogramma* wasps that are released weekly from mid December until the end of February. Monitoring of MNB activity dictates the wasp release areas.

Doug is totally committed to using IPM. His experience has shown that it is the most cost-effective way of controlling pests, while at the same time minimising environmental impacts.

Macadamia nuts (2): Andrew and Jacqueline Heap, Wollongbar, New South Wales

Andrew and Jacqueline Heap own and manage an 8-hectare macadamia nut farm near Wollongbar on the North Coast of New South Wales. The farm consists of 1800 macadamia nut trees, which produce between 19 and 30 tonnes of nuts, depending on the year.

Until the 2010–11 season, control of insect pests was achieved with the use of insecticides alone. Several factors contributed to the decision to try something different, including difficulties in spraying in wet conditions, and also the fact that the area is becoming more densely populated, which can make the use of insecticides more difficult. Following the completion of a biological farming course at a nearby TAFE college, and after many discussions with other farmers, Jacqueline decided to use an IPM approach, which involved minimal use of insecticides and release of parasites (*Trichogramma* wasps) of macadamia nut borer.

The farm is relatively small, so it was decided that the new approach would apply to the entire farm, not just a section. Finding out about availability and use of *Trichogramma* wasps, (where to buy them, how to release them, how many were needed and how long they could be stored) was the first step. However, biological control agents cannot be bought for all of the pests, so integrating control measures (biological and chemicals) also needed to be worked out.

Although the *Trichogramma* wasps may be more expensive than the amount of insecticides that were used in the past, it is actually cheaper when taking into account the fuel and labour to apply two or three sprays. There are also advantages in reduced soil compaction, because tractors are not driving down the rows so often, and also the workers are happy to staple a few strips of paper (containing the *Trichogramma*) to a tree rather than have to handle insecticides. In the last season it was extremely wet, and Jacqueline still achieved control of the pests when it would have been impossible to drive a tractor through the property. The main losses were from fruit-spotting bug, as there was no biological control available and sprays could not be applied.

The decisions on what actions to take are now based on monitoring of insect numbers and damage to nuts. The availability of chemicals and contractors to apply them is also a factor that is considered, as well as ease of management and worker safety and availability.

Onions: Harvest Moon, Tasmania

Harvest Moon grows a wide range of fresh-market vegetables, with the main lines being broccoli, onions, potatoes and beans, but also other crops, such as carrots, celery and lettuce. The company produces around 50 000 tonnes of produce per year on approximately 1600 hectares of farmland on Tasmania's north coast and central district. Approximately half of the production comes from its own farms, with local contract farmers supplying the remainder.

The company first became interested in IPM some years ago when it trialled an IPM approach in 8 hectares of broccoli production. The success there prompted it to look at an IPM approach in some of the other crops.

Mark Kable, Agricultural Director for Harvest Moon, says that the members of the company are impressed with what has been achieved using IPM, and plans to implement it in all their vegetable production. 'We are currently using IPM for our broccoli, lettuce and onions, using less insecticide and getting better control

Figure 8.2 Harvest Moon Agricultural Director Mark Kable (left) and Kevin Temple. (Photo courtesy Harvest Moon)

of insects and other pests.' Although Harvest Moon was initially sceptical, Mark now believes that IPM strategies can be used in all of its crops, and is keen to see this happen.

> We have changed our whole agronomy structure due to our involvement with Paul Horne, Jessica Page and IPM. There is now much greater emphasis on scouting and monitoring of pests and diseases. We are now much more aware of the role of beneficial insects in an intensive cropping regime.

Mark also noted that 'IPM is a very important part of our business. It not only saves us money from reduced chemical inputs, but it also makes us think more about other issues such as beneficial species, weeds in neighbouring paddocks that harbor certain insects and general good farming practices.'

Organic onions and asparagus: Maurie and Maria Cafra, Koo Wee Rup, Victoria

Maurie and Maria Cafra run an organic farm, (certified with NAASA), growing a range of vegetable crops near Koo Wee Rup in Victoria. The farm is about 70 hectares in size, and approximately half the area is used to grow asparagus, 10 hectares is devoted to onions, and a range of other vegetables, including tomatoes, eggplants, capsicums, brassicas, corn, green beans and artichokes, account for the remainder. At first Maurie grew crops, as his father had done, with pesticides as the basis of pest control, but in 1997 Maurie converted his farm to organics, and at the same time decided to use IPM strategies to deal with the range of pests present. So the changes that Maurie made to pest management methods were massive.

The initial reason for the change was simply to avoid spraying conventional pesticides. However, Maurie wanted not just to stop spraying but actually implement better pest management methods. It began with working on control of aphids in brassicas, then controlling redlegged earth mite in asparagus. After trying some organically allowable (but non-selective) sprays, such as lime sulphur and pyrethrum, Maurie put more effort into cultural controls and building up biological controls. The only pesticides used now are products that have no detrimental effects on beneficial species (such as Dipel, a bacteria).

Maurie noticed a shift in the types of pests and the severity of pests over about a 5-year period. For example, diamondback moth in brassicas became less important and cabbage white butterfly became more important. Beneficial species and better weed control meant that redlegged earth mites and aphids stopped being serious problems at all.

Weed control remains a bigger problem than dealing with insects and mites now, as Maurie is confident that the ecological change that has occurred on his farm has resulted in a much better balance between pests and beneficial species.

Ornamental nursery plants: Evergreen Nurseries, Victoria

Michael Kelly and his son Ben run Evergreen Nurseries, which is a company that grows about 200 different types of ornamental plants for wholesale markets, and in particular concentrates on climbing plants and *Fuchsia*. Each year they produce between 225 000 and 250 000 plants for sale.

In 2003 Michael was concerned that he was having a contract sprayer come and spray his whole nursery regularly for insect and mite pests, yet he would have to do it again himself in between visits. That is, the pesticide-based strategy for controlling pests was failing, mainly because of the development of insecticide resistance. Michael did this extra spraying himself on weekends to avoid times when staff were present, but in addition to the loss of several hours he found that he often had headaches afterwards. In part this could have been related to increasing use of what he considered to be more potent pesticides in order to deal with the tougher pests.

In an endeavour to improve the situation, full blanket spraying of the whole nursery was discontinued, and for several years a system of spot-spraying outbreaks of insect pests with so-called 'greener' alternatives, such as chilli and garlic sprays and pest oils, led to some improvements.

That prompted Michael to try to use beneficial insects and mites as part of an IPM program. However, his first attempts failed and he went back to spraying. He later realised that he had not had enough information about the impact of different pesticides (including fungicides) on the beneficial species that he was using. Basically he had been buying and releasing beneficial insects and mites, and then killing them.

In 2008 Michael and Ben contacted Paul Horne and Jessica Page to try once again to use a different approach, but this time with more precise information. The main trigger for deciding to make some changes was the problem that they were having with western flower thrips, in addition to the pests that they had dealt with in the past. Michael and Ben felt that they needed an IPM expert to help guide them in person with specific problems as they arose. They decided to make the change to IPM on their whole farm rather than just part of it, mainly because they wanted a consistent approach across the farm. This was to avoid having different rules on pesticides in different sections of the farm.

This time they worked out how long they had to wait before introducing different species of beneficials (because of pesticide residues) and how long before they were likely to gain control. This allowed them to work out the best time of year to commence the new program, taking into account both the main season for sales and the time of greatest pest pressure.

Progress was made quickly, but without achieving total control in the first season. In the second season of using the IPM approach they found great success, as measured by the improved quality of their main crop (*Fuchsias*) because of better control of western flower thrips. Control of western flower thrips (the main reason for implementing IPM) has been successful, but has taken 2 years before reaching the stage of full control in terms of 'no economic loss' not 'no thrips present'. The problems that now need to be dealt with at Evergreen Nurseries are relatively minor pests that were formerly controlled by broad-spectrum insecticides. For example, lightbrown apple moth is a pest that was previously not a problem, but now needs to be recognised, considered and dealt with as it appears. So, although the pests can be dealt with and are only a minor concern in production, the problem is that buyers are demanding a product with no damage whatsoever. Controlling an insecticide-resistant pest such as western flower thrips is the most important thing to Michael and Ben, but damage by any insect, however easy to control, is of concern to buyers. Michael and Ben do not want to disrupt control of the really serious pests by spraying an insecticide targeting a minor pest, but need to control all pests adequately to the satisfaction of their customers.

The difficulty that Michael and Ben now face is having to recognise all the different life stages of a range of beneficial species as well as pests, and deciding on pesticide applications where necessary. This is where they call on external advice. Michael says, 'We have many books and photos, but in practice we need to learn how to be confident in our monitoring and decision-making, and that is where we need expert assistance.'

The advantages of the changes that have been made include better control of key pests that previously could not be controlled with ever-increasing pesticide applications, but also a much better feeling about the minimal use of pesticides. This is partly because Michael's house is in the middle of the nursery, but there is a much better Occupational Health and Safety aspect to their approach that both staff and owners appreciate. Michael comments that since making the changes to adopting IPM and minimal pesticide use he has seen frogs and magpies return to the farm. This is a side effect that indicates the safer aspect of their new approach.

Potatoes: Wayne Tymensen, Victoria

Wayne Tymensen is a potato grower in Gippsland, Victoria. He grows 120 hectares of potatoes each year for the crisping potato industry. He first encountered the possibility of using a different approach to pest management (an IPM approach) in 1995, and thought that it was worth a trial. His next-door neighbour (Peter O'Sullivan) agreed, so both farmers decided to give IPM a go. Wayne began by trialling an IPM approach on two paddocks, each about 4 hectares. Within 2 years he was using IPM on the entire farm. Wayne has now bought out his neighbour, so the farms have effectively become one, and IPM is the basis of pest control.

Before 1995 Wayne and his father dealt with pests using a regular pesticide-based program with a broad-spectrum insecticide applied approximately every 2 to 3 weeks, but more often if considered necessary. They rotated through a variety of chemical groups, using what was generally considered best practice at the time. The idea of using a different approach based on minimal, possibly no, insecticide applications was of interest because Wayne wanted a more sustainable method of dealing with pests, but there was caution in case it did not work. So they decided to give it a try on two paddocks, but certainly not the entire farm. The results were good, so they cautiously tried it on more paddocks in the following year. With more positive results in the second year they were encouraged to adopt the new strategy based on minimal pesticide use on the entire farm. Adopting a new approach required on-farm trials and several years. It was not only Wayne that had to learn. His IPM adviser, Paul Horne, also had to learn how to apply his entomology experience into a regular commercial advisory service for potatoes, so it was a collaboration between farmer and entomologist that allowed IPM adoption to occur. After 16 years it is still going.

No insecticide is applied to Wayne's potato crops unless there is a specific reason for doing so, and then the most highly specific pesticide available is chosen. In most years no insecticide is recommended or applied on any paddock. The change that Wayne has seen on his own farm in that time has been from applying insecticides routinely several times in each crop to perhaps once (usually none) per crop per season. This has not only given Wayne a reduced pesticide bill but also resulted in better control of pests. This second result is counterintuitive, as previously Wayne would have thought that applying insecticides would always reduce pest numbers. He has seen the opposite occur, and now would like to encourage his neighbours to stop spraying routine insecticides as that would reduce the pest pressure on his crops. Control of pests, especially potato tuber moth, has been greatly improved since using IPM.

Disruption of beneficials by pesticides has demonstrated to Wayne just how important the role of beneficial species is in his crops, and how careful he needs to be in selecting pesticides (see Figure 8.3). In early 2010 Wayne called for a plane to apply a fungicide (called 'Ace'). However, a mistake was made, and half of the crop was sprayed with an insecticide (called 'Axe'). Weekly monitoring of aphids in particular showed the difference in aphid counts in the two sections of the paddock. Aphid numbers rose in the section sprayed with 'Axe' because it killed the predators and parasitoids that would have eaten them (which is what occurred in the other half of the crop where there had been no disruption).

Seeing this result has emphasised to Wayne just how powerful a tool the biological control agents are in his crops. In the early days, not applying an insecticide was difficult when pests were around. Wayne says, 'The only disadvantage is you can feel helpless and a little on edge when the insect pressure is at its greatest; for example, when you have some very hot weather and the potato tuber moth numbers get very high with a crop of potatoes destined for storage. When applying an insecticide it feels like you have done something.'

Now, if concerned about anything regarding insects or insect-vectored diseases, Wayne calls his IPM adviser (Paul Horne) to discuss the potential

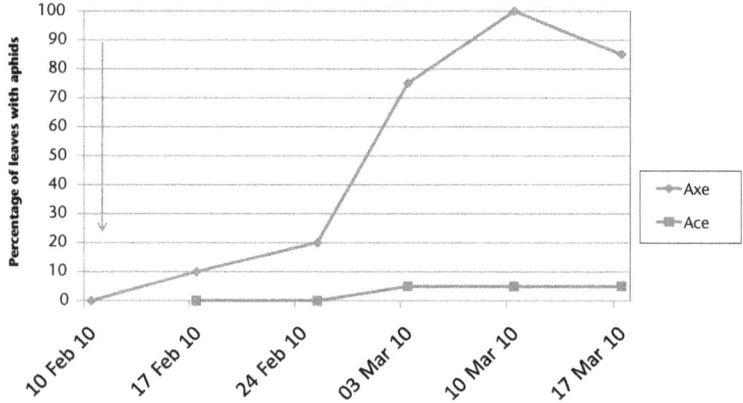

Figure 8.3 *Aphid numbers* in sections of the potato paddock sprayed with *Axe* (a synthetic pyrethroid) by mistake (indicated by arrow), versus the section sprayed with fungicide only (*Ace*)

problems and the options available for control. Given the length of time that Wayne has been using IPM, he is obviously sure of the advantages, but remembers how he felt in the first few years. He says, 'I think if someone is unsure about IPM but would like to give it a go, try a paddock, and I would be very surprised if you didn't change over to full IPM.'

Roses (hydroponic flowers): Boon Roses, Officer, Victoria

Boon Roses <http://www.boonroses.com.au> grows hydroponic roses on two properties near Officer in Victoria. Boon Roses is among the largest of the glasshouse rose farms in Victoria. It has 5 hectares of glasshouses growing the highest quality cut roses, producing between 5 and 6 million stems per year.

On one of these two farms Willem Boon and Debra Dawson are responsible for the production of around 350 000 bunches, or 3.5 million stems of roses annually. In 2000, Boon Roses was looking at costs for controlling the (then) major pest, two-spotted mite, and the decreasing degree of control that was achieved. Willem and Debra wanted to know if this pest could be controlled. We said 'Yes (easily)', but we would need to look at control of all pests, not just one. Then Boon Roses was using a strategy that relied totally on chemical pesticides.

The reason for even trying anything different was that spraying for mites was becoming less and less effective. Willem and Debra had tried to use *Persimilis*, the predatory mite that eats two-spotted mite, but without success. They now understand that the pesticides they were using for control of other pests and diseases were not compatible with *Persimilis*, and it was certain to fail.

However, since then western flower thrips has become the most significant pest, surpassing two-spotted mite. This pest can also be dealt with using an IPM approach. This strategy has been implemented at Boons Roses, but it needs to be a part of an overall approach or control of two-spotted mite will be lost. There is no advantage in controlling one pest but in the process losing control of another. Previous minor pests, such as aphids and mealybugs, have become more serious concerns on the farm as they cannot be treated with the previously used insecticides or control of western flower thrips and two-spotted mite will be disrupted.

The change to using an IPM approach based on biological control was implemented in stages, in one house at a time. It commenced in an area being reestablished with new plants, then applied through the older plants. Some of the pesticides that had been sprayed have a residual effect, killing *Persimilis* and other beneficials for 12 weeks or more. So the initial stage involved in changing over had to deal with the pests without adding to the residue time, effectively weaning off the pesticides until the beneficial species could work. Willem and Debra found this initial stage, waiting for the beneficials to establish and achieve control, the most difficult part.

Information and training on how to recognise the different life stages of the different species of beneficials was essential, but the most important information that Willem and Debra needed was about the effects of pesticides on each species of beneficial; for example, that Pirimor (pirimicarb) will kill some species of beneficials but is not very residual, but other pesticides, such as Regent (fipronil) will kill all beneficials that they want to use for 3 to 4 months.

Once they saw the new strategy working they became more confident, and have become aware of several advantages. The first is the reduced use of insecticides (now a rare event) and treating only small areas when necessary, not whole houses or the entire farm. Another unexpected advantage was the increased vase life of the cut roses and improved health of the plants. From a management point of view, the IPM approach is more difficult than using a regular spray program, but is worth it according to Willem and Debra. The advantages in crop quality far outweigh the difficulties of monitoring and decision-making. They note that the initial major problem that prompted the change to IPM, two-spotted mite, is actually the easiest to control. Willem comments, 'If you can't control two-spotted mite with *Persimilis* then you can't do IPM in roses'.

Strawberry crops: Wombat Berries, Wandin, Victoria
Wombat Berries is a family farm run by Eddie DiPietro, his daughter Clarissa and son-in law Bobbie Cincotti, and the decisions on what pest management options should be taken differ from those described above. The farm consists of 30 acres (12 hectares) near Wandin, Victoria, where they grow about 300 000 strawberry plants each year.

Wombat Berries had previously used predatory mites to control two-spotted mite, but with only limited success. The main method of controlling insect and mite pests was with applications of pesticides. When the pesticides that they had been relying on stopped delivering results (the sprays did not adequately control the target pests), they were interested in trialling a different approach and changing their approach to pest management if necessary. The fact that the pesticides had stopped working, along with the concerns about consumer perceptions of what was sprayed on strawberries, made the people at Wombat Berries consider an alternative approach.

The information that Wombat Berries needed in order to implement IPM was not readily available. They wanted information on identification of insects (pest and beneficial), lifecycle information on both pests and beneficials and advice on the impact on a range of pesticides that they could choose to spray or not.

What Wombat Berries needed was expert IPM advice, especially as to what was the impact of each pesticide on each species of beneficial invertebrate. This information had not previously been available, but IPM Technologies (Paul Horne and Jessica Page) were able to provide such advice when required. This meant that Wombat Berries had an IPM advisory service when needed.

The first thing that was suggested was that control of two-spotted mite was essential before control of other pests (including western flower thrips (WFT)) was attempted. The predatory mite *Persimilis* needed to have controlled two-spotted mite so that miticide sprays became no longer necessary, as the main commercially available predator for western flower thrips is a mite.

The path to achieving this has not been easy. At several times during the growing season Wombat Berries was suffering significant losses because of WFT. The option to return to insecticide sprays to deal with this pest was considered but rejected. The owners of Wombat Berries knew that there was at best a very short-term advantage to be had and that they should look to the longer term. They decided to rely on biological control agents (both naturally occurring and commercially available) and watched the beneficial species control the pests while on neighbouring farms the sprays failed.

Clarissa notes that some of the changes in management have taken time to get used to, including the fact that a pest population will always be present, and it is necessary to be patient and resist the urge to use chemicals too soon for a short-term solution. Eddie adds that they have needed to change their previous (long-held) attitude that chemicals are the solution to all pest problems. Also, 'When you have an insect problem and you physically go out with the tractor and apply a chemical to the strawberries then you feel as though you have done something to stop the problem. When you use IPM, at first it sometimes feels as though you are sitting on your hands and doing nothing about the problem. You don't go home at night feeling that you have done something to prevent a loss in income and fruit quality.'

They now know that pesticide applications can often make the problems with WFT and mites worse, and now they spray far less insecticide and enjoy the fact that they need to spend less time spraying. Chemical use is now a last resort, and based on the results of monitoring of both pest and beneficial species.

Regular monitoring of thrips in particular allows them to compare what is happening on their own farm, and in individual blocks, with other farms. This means that they can assess the factors, such as hot weather, leaf-trimming, release of predators and chemical applications, that influence both pest and beneficial populations.

Wine grapes: Rising Vineyard Neil and Elizabeth Roberts, St Andrews, Victoria

Neil and Elizabeth Roberts grow premium quality grapes on their 32-hectare vineyard in St Andrews, Victoria, which is the largest in the Diamond Valley. The Diamond Valley is adjacent to the better known Yarra Valley near Melbourne, Victoria, Australia.

In 1996 they decided to trial a different approach to dealing with pests (IPM). Until that time the approach to dealing with pest insects and mites had been to apply an appropriately registered insecticide or miticide. Then they began to appreciate that not only could certain pesticides be ineffective but that they could also induce pest problems.

The questions that we asked Neil and Elizabeth were the same as for everyone else reported in this book, but in this case the responses are exactly as received.

Changing to using IPM

1   Background: Size of farm; what do you grow?; how many bunches/tonnes, etc?
    This vineyard was started (as a part-time venture) in 1979. No good advice, relevant to local conditions, was available at the time as the replanting of the Yarra Valley vineyards had then just commenced. The property was gradually expanded, and planting area increased over the years that followed. Planted area has been 80 acres (32 ha) since 2000, and in the years since we became full-time professional growers we have progressively redeveloped and replanted the early vineyard blocks. Replanting still continues under a program of continuous improvement commenced in 2001. We produce wine grapes for the premium wine sector, with most of the crop contracted to Coldstream Hills, the premium Yarra Valley label of Treasury Wine Estates. Production is typically between 220 and 250 tonnes p.a.

2   How did you control pests before using IPM?
    Prior to 1996 we controlled pests reactively using a broad-spectrum insecticide (such as Lorsban). The principal pest was (and continues to be) LBAM (lightbrown apple moth).

3   What made you want to try something different?
    We attended a presentation on IPM presented by Dr Paul Horne about 1996, and the ideas resonated with us as a viable alternative to the regular use of broad-spectrum insecticides. We sensed that the industry would, at some time in the future, be looking to promote environmentally sensitive and sustainable practices in the vineyard, and we were keen to be at the vanguard of that change. This has proven to be the case, and has been part of our 'competitive advantage' as the 'clean and green' movement has gained momentum.

4   Did you try to change the whole farm at once, or bit by bit?
    We did not see the progressive change option to be viable, as IPM requires the farmer to promote the health and viability of the predator insect population. This would not occur unless *no* broad-spectrum insecticides were to be used. We adopted the total and abrupt changeover model and began to have the problem and predator insect population professionally monitored. This practice continues to this day, with an average professional monitoring period of 10–12 weeks per annum the norm.

5   What was the first step?
    Total cessation of broad-spectrum insecticide use; employment of consultant
    entomologist for monitoring; accurate recording of pest–predator numbers.
6   What information and advice was necessary?
    IPM Technologies Pty Ltd has been retained for the advisory/monitoring task
    from day 1, and each year since.
7   What are the problems and disadvantages?
    Professional monitoring and advice involve the expense of the consulting fee.
    They also involve some time in liaising with the consultant and the discussion
    of treatment and non-treatment options.
8   What is the advantage of IPM now?
    We have an accurate and scientifically based assessment of pest species
    numbers and an assessment of the capacity of the predator insects' capacity to
    deal with the numbers emerging. This is a clear contrast with the typically
    anecdotal and reactive process that typifies farmers' pest insect strategies.
9   What has changed since you first started?
    We have learned how to refine the pest numbers tolerances at various stages in
    the season to minimise both insecticide use and damage impact.
10 How do you make decisions on pest management?
    The consultant provides an email report and the opportunity to discuss
    options for treatment.
11 Anything else you think is important.
    Rigorous monitoring is the key to successful implementation of IPM concepts.
    This *can* be done by the farmer, but in my experience is unlikely to be done
    with the regularity and rigour required. While 'arms length' professional
    monitoring is expensive, it should provide the basis for more objective decision
    making and avoid the haphazard and irregular monitoring that might
    otherwise occur. The farmer needs to establish a relationship with his
    monitoring agent that is open and trusting, so that a good two-way
    communication can underlie the IPM implementation.

Chemical resellers and agronomists (1): Serve-Ag Pty Ltd, Tasmania
(This text was part of an article that was published in *Good Fruit and Vegetables*
magazine in 2010.)

Serve-Ag Pty Ltd has been providing production advice to fruit and vegetable
growers in Tasmania for over 30 years. During this time, advice from Serve-Ag
agronomists for the management of pest and diseases has been very important for
the profitability of growers. Agronomist Peter Aird has worked with Serve-Ag for
over 25 years, and is very keen for agronomists to use an IPM approach more often
in a whole range of situations. 'We have seen IPM work well in vegetable crops, and

with help from Paul Horne and Jessica Page we have also applied the same approach successfully in other crops, such as grapevines, cereals and canola.'

Peter remembers meeting Paul and Jessica at a demonstration of IPM at Forthside Research Station near Devonport following the arrival of lettuce aphid in Tasmania. 'Halfway through the life of the crop there were high levels of lettuce aphid in all the lettuces and I thought that there was no way that the IPM demonstration would work. However, at harvest the predatory insects had eaten the pests, and there was a totally clean crop. That field trial was a powerful demonstration of what IPM could achieve.'

Peter comments, 'Initially, there was a reluctance to adopt all the recommendations from IPM Technologies because of the requirement of some industries QA procedures having a zero insect tolerance, but after discussion with grower groups, field officers and industry, it became apparent that there was a need to be more flexible in post-harvest assessments. After assessing early results, insect management techniques using an IPM approach were equal or better than the standard practices.'

Graeme Palmer (Agricultural Production Manager) wants to give young Serve-Ag agronomists more training in IPM so that they are better able to provide the sort of advice that is now being requested by farmers in Tasmania. 'Involvement in this project has improved Serve-Ag's pest management knowledge. The project has improved Serve-Ag agronomists' identification of beneficial insects and assisted with increasing their knowledge of the behaviour and habitat required to best support beneficial insects.' Rebecca Clarkson is one new agronomist who is finding the IPM approach interesting, and she wants to use this type of management more often.

Graeme points out, 'The Tasmanian vegetable and fruit industries are now investigating other pest management issues with Serve-Ag regarding an IPM approach to dealing with pests. These include guidelines for control of slugs in vegetables and assessment criteria for white fringed weevil management, and also control of pests in pome and stone fruit.'

Chemical resellers and agronomists (2): John Frisina, Landmark, Wandin, Victoria

John Frisina is the branch manager of Landmark in Wandin, Victoria. Landmark, Wandin, services growers of a range of horticultural crops, including glasshouse flowers, tree fruit (apples and pears), vineyards and berry fruits, and provides pesticides, fertilisers and other products as required. Landmark provides assistance to growers with pest control issues, and supports growers in whatever way they would like to deal with pest issues. In most cases up until the last few years this has meant supplying the most appropriate pesticide, but has also involved supplying predatory mites.

Although the main demand from growers has been for suitable pesticides, John Frisina and his team have a long history of providing other, IPM-type support, where growers have requested that type of product or advice. Apple growers, flower growers, vineyard managers and berry growers in his region have all used commercially produced biocontrol agents (*Persimilis*) to control two-spotted mite, and Landmark has been able to provide these products instead of just pesticides. The difficulty for both growers and Landmark has been to know which (pesticide) products are compatible with the release of predators.

Two-spotted mite has long been the most important pest, but in recent years the control (or lack of it) of western flower thrips has become more important than dealing with any other pest. For a few years the insecticide spinosad (Success) did effectively control WFT, but of course the pest with a very short generation time developed resistance to this product. It had also developed resistance to almost every pesticide available. This situation was obviously extremely serious for the local strawberry industry, and led to a crisis meeting in 2008 held in the offices of the Department of Primary Industries, Knoxfield. John was the facilitator, and the strawberry industry decided to explore the possibility of an IPM approach to dealing with this pest.

The change to using an IPM approach in strawberry crops has been difficult, and has led to Landmark staff having to deal with growers in a slightly different way. The first step is to determine whether or not clients wish to use IPM based on predatory insects and mites, then to determine what options are available. Instead of simply supplying a pesticide, Landmark staff need to ask a series of questions about what actions have been taken, what pesticides have been sprayed or beneficial species have been released and when, and what actions are likely in the next weeks. Even the species of beneficials and the particular pesticides that may be used are required information.

Landmark is also able to order and supply beneficial insects and mites that are available from commercial producers, so they need to know about the effectiveness and limitations of the biological control agents. All of these new issues mean that Landmark staff who are providing advice to farmers need to be aware of all options and impacts of all possible recommendations, which is now much more complicated than in previous years. John noted that information sessions on IPM often prompted more questions on what was required than were originally answered.

Chemical resellers (3): Jerome Thompson, Werribee South Farm Supplies, Werribee South, Victoria

Jerome Thompson is branch manager of Werribee South Farm Supplies (a CRT store). He supplies agricultural products, including pesticides and advice on pesticides, to about 110 vegetable farmers in the Werribee South region of Victoria.

Over the last 10 years Jerome has seen a progressive change in farmers' use of insecticides in particular and also their attitude towards IPM.

Previously Jerome would mainly sell products such as Endosulfan, Lannate, Nitofol, Folidol, Dominex and Rogor, but issues such as insecticide resistance, problems with the older chemicals in terms of safety, having to handle 20-litre drums, chemical odour in the store and knowing that there were newer, more selective chemicals available prompted him to talk to growers about using a different approach. He did this by discussing other options, and in particular the use of IPM, with a couple of growers at a time, and seeing how they responded and the results that they achieved.

Initially Jerome encouraged growers to use products based on *Bacillus thuringiensis* (B.t.) and importantly, not to mix them with the older, broad-spectrum insecticides. However, growers felt they needed to have a 'safety net' in case the B.t. failed, so the changed approach was progressive. What was necessary was for growers first, to get better at monitoring pests and timing sprays; but also to start recognising beneficial species and monitoring for those as well. This was a big change for growers; identifying beneficials in the crops is still the biggest issue to overcome.

Another difficulty Jerome encountered when encouraging growers to use B.t. products was getting them to spray late in the day to avoid UV degradation. However, growers' attitudes to spraying practices have largely changed, and now there is much less concern about insecticide resistance and having to handle the old products.

The lack of chemical odour in the store is a very noticeable change, because stocks of the older broad-spectrum insecticides are kept at an absolute minimum (one drum each of three products). The laws governing the storage and handling of 'dangerous goods' and especially S7 products, are strict, and so not having stocks of such products is a bonus for Jerome as manager of the store. Now the insecticides used are almost all selective products, and Jerome encourages farmers to use them within an IPM framework rather than stand-alone products.

Chemical companies: for example, DuPont, Bayer, Sumitomo

The companies that produce pesticides that are currently demanded by farmers have also adopted massive changes in the last decade. This segment of the book is not something that the companies referred to have vetted or endorsed, but this is our own opinion of the challenges they are facing and their recent responses to those challenges.

They face a difficult situation, as they may produce and sell both broad-spectrum insecticides (often in great demand by farmers) and the newer, more selective and more expensive insecticides that can have a place within IPM strategies. The onus is on the farmer to ask for what he or she wants: either a cheaper, old-style broad-spectrum insecticide or a selective insecticide that has an effect on only certain species of beneficials.

The newer products are certainly more IPM compatible, as they do not kill all species of beneficials. What is needed is an assessment as to what species of beneficials are killed, and whether these species are important in particular crops. As we have described in previous chapters, the decision as to whether or not an insecticide is 'safe' or not depends on the species that is present in any particular crops. So the promotion of the newer insecticides that definitely are not broad spectrum still depends on a correct assessment of what the product is safe on and what it will kill or affect.

In recent years the major chemical companies have been interested in promoting some of their products as being more compatible with IPM strategies where it may be appropriate. There are often different opinions on what defines IPM (see Chapters 1 and 6), and even more differences of opinion on what is a 'safe' pesticide in terms of beneficial species. So chemical companies such as DuPont, Sumitomo and Bayer (among others) have contracted research by a range of organisations, including our own IPM Technologies Pty Ltd, to find out the effects that some of the newer actives or formulations will have on particular beneficial species in crops where the product is to be used.

Information of this type is only of value to farmers who want to change (or have changed) from a pesticide-based strategy to one that involves both beneficial species and selective pesticides. It shows how the chemical companies are also changing in response to the changing requirements of farmers. (Some of the results of this type of research can be seen on websites such as those of IPM Technologies Pty Ltd, Koppert and BioBest). Once again, it is collaboration between IPM entomologists and chemical companies that will get the best result in terms of best use of pesticides.

In Australia, at the annual vegetable industry conference in Brisbane in 2011, a representative from Bayer presented exactly this viewpoint. Bayer was divesting itself of products that were broad-spectrum and concentrating only on selective products that had a potential fit in IPM systems.

A final point

Whatever the issue it is difficult for most people to make a change in established practices. However, it is entirely possible for farmers (and advisers) to change from using a pesticide-based approach to using an IPM strategy. What is required is a collaboration between farmers and their advisers (including entomologists and their representative industry bodies). For farmers wanting to make such a change, help is available.

Help is available in the form of entomologists (although they can be hard to find), manuals and on-line support. The availability of digital cameras, mobile phones and emails make timely advice more possible than ever before. Care is needed in interpreting so-called IPM information that is broad-brush rather than specific. At present there is more general, non-specific information available than highly specific information, and this is highly likely to lead to failure in adoption of IPM. So we suggest that anyone interested in adopting a changed approach to pest management be careful about the type of information that is used and insist on highly specific information.

It is not necessary or always desirable to change from a familiar pest-control strategy to something entirely different. Some farmers will choose to make the decision to change everything over the entire farm immediately (for example, see the section on wine grape production, pp. 99–101), while others will try a paddock or two before the whole farm (for example, see the section on arable crops in New Zealand, pp. 83–84). We suggest that you seek advice from specialists in IPM or any other aspect of change that is required. Push the agronomists to deliver the type of pest-management advice that you are seeking, and do not accept the generic pesticide-based recommendations!

It works!!!

Glossary

APVMA   Australian Pesticide and Veterinary Medicine Authority

B.t.   *Bacillus thuringiensis*, the bacteria used in products such as Dipel

Confidor   An insecticide in the neonicatinoid group, active ingredient imidacloprid

Dipel   An insecticide based on the bacterium *Bacillus thuringiensis* (also known as B.t.)

Honeydew   The sugary waste excreted by aphids and other sap-sucking insects

IPM   Integrated Pest Management

IRM   Insecticide Resistance Management

Movento   A true systemic insecticide; active ingredient spirotetromat

MRL   Maximum Residue Level

population dynamics   The study of how populations of different species interact (such as predator and prey, or two species of pests competing for the same host plant)

secondary poisoning   Poisoning of a predator or scavenger that eats prey that has been killed by pesticides

TAFE   *Tertiary And Further Education:* An education option that provides a range of tertiary education for farmers and their employees

withholding period   The minimum time interval that by law must be observed from the time that a chemical is sprayed until the time of harvest. This varies between products and crops

Common and scientific names of species mentioned in this book

Balaustium mite   *Balaustium medicagoense*

bald eagle   *Haliaeetus leucocephalus*

black field cricket   *Teleogryllus commodus*

blue oat mite   *Penthaleus major*

brown lacewing   *Micromus tasmaniae*

Bryobia mite   *Bryobia rubrioculus*

cabbage white butterfly   *Pieris rapae*

cane grubs   species of family Scarabaeidae: sub-family Melolonthinae

cane toad   *Bufo marinus*

cape weed   *Arctotheca calendula*

carabid beetles   beetles in the family Carabidae (ground beetles)

Carpophilus beetle   *Carpophilus* spp.

Citrophilus mealybug   *Pseudococcus calceolariae*

cocoa pod borer   *Conopomorpha cramerella*

codling moth   *Cydia pomonella*

corn earworm   *Helicoverpa armigera*

cottony cushion scale   *Icerya purchasii*

damsel bugs   *Nabis kinbergii*

diamondback moth, *Plutella*   *Plutella xylostella*

Encarsia   *Encarsia formosa*

green lacewing   *Mallada signatus*

Heliothis (armigera)   *Helicoverpa armigera*

Heliothis (punctigera)   *Helicoverpa punctigera*

house fly   *Musca domestica*

Indian mynah   *Acridotheres tristis*

ladybirds   Family Coccinellidae

lettuce aphid   *Nasonovia ribis-nigri*

lightbrown apple moth   *Epiphyas postvittana*

lucerne flea   *Sminthuris viridis*

Mite B   *Neoseulius wearnii*

native budworm   *Helicoverpa punctigera*

***Persimilis*, Chilean predatory mite**   *Phytoseulius persimilis*

plague thrips   *Thrips imaginis*

potato tuber moth   *Phthorimaea operculella*

prickly pear   *Opuntia graminis*

redlegged earth mite   *Halotydeus destructor*

Rutherglen bug   *Nysius vinitor*

spider mite   *Tetranychus urticae*

sow-thistle aphid   *Uroleucon sonchii*

spotted amber ladybird   *Hippodamia variegata*

tachinid flies   order Diptera: family Tachinidae

transverse ladybird   *Coccinella transversalis*

tubular black thrips   *Haplothrips victoriensis*

two-spotted mite   *Tetranychus urticae*

vedalia ladybird   *Rodolia cardinalis*

wasps   Hymenoptera

western flower thrips   *Frankliniella occidentalis*

white fringed weevil   *Naupactus leucoloma*

Species mentioned, listed by scientific name

Acridotheres tristis   Indian mynah

Arctotheca calendula   cape weed

Balaustium medicagoense   *Balaustium* mite

Bufo marinus   cane toad

Bryobia rubrioculus   *Bryobia* mite

Carpophilus **spp.**   *Carpophilus* beetle

Coccinella transversalis   transverse ladybird

Coccinellidae   ladybirds

Encarsia formosa   *Encarsia*

Frankliniella occidentalis   western flower thrips

Halotydeus destructor   redlegged earth mite

Haplothrips victoriensis   tubular black thrips

Helicoverpa armigera   *Heliothis*, corn earworm

Helicoverpa punctigera   *Heliothis*, native budworm

Hippodamia variegata   spotted amber ladybird

Hymenoptera   wasps

Icerya purchasii   cottony-cushion scale

Micromus tasmaniae   brown lacewing

Musca domestica   house fly

Nabis kinbergii   damsel bugs

Nasonovia ribis-nigri   lettuce aphid

Naupactus leucoloma   white fringed weevil

Neoseulius wearnii   Mite B

Nysius vinitor   Rutherglen bug

Opuntia graminis   prickly pear

order Diptera: family Tachinidae   *Tachinid* flies

Penthaleus major   Blue oat mite

Phytoseulius persimilis   *Persimilis*, Chilean predatory mite

Pieris rapae   cabbage white butterfly

Plutella xylostella   diamondback moth, *Plutella*

Phthorimaea operculella   potato tuber moth

Rodolia cardinalis   vedalia ladybird

Sminthuris viridis   lucerne flea

Thrips imaginis   plague thrips

Tetranychus urticae   two-spotted mite

Uroleucon sonchii   sow thistle aphid

References

Bailey P and Caon G (1986) Predation on two-spotted mite, *Tetranychus urticae* Koch (Acarina, Tetranychidae) by *Haplothrips victoriensis* Bagnall (Thysanoptera, Phlaeothripidae) and *Stethorus nigripes* Kapur (Coleoptera, Coccinellidae) on seed lucerne crops in South-Australia. *Australian Journal of Zoology* **34**: 515–525.

Bajwa WI and Kogan M (2003) IPM adoption by the global community. In *Integrated Pest Management in the Global Arena*. (Eds KM Maredia, D Dakouo and D Mota-Sanchez) pp. 97–107. CABI Publishing, London.

Carson R (1962) *Silent Spring*. Houghton Mifflin, New York.

Horne PA and Page J (2008) *Integrated Pest Management for Crops and Pastures*. Landlinks Press, Melbourne.

Horne PA and Page J (2009) Control of western flower thrips (*Frankliniella occidentalis*) in lettuce and strawberry crops in Victoria, Australia. *International Symposium on Thrips and Tospoviruses*, Brisbane 2009.

Horne PA, Page J and Nicholson C (2008) When will IPM strategies be adopted? An example of development and implementation of IPM strategies in cropping systems. *Australian Journal of Experimental Agriculture* **48**: 1601–1607.

Horrocks A, Davidson MM, Teulon DAJ and Horne PA (2010) Demonstrating an integrated pest management strategy in autumn-sown wheat to arable farmers. *New Zealand Plant Protection* **63**: 47–54.

Smith D and Papacek DF (1991) Studies of the predatory mite *Amblyseius victoriensis* (Acarina: Phytoseiidae) in citrus orchards in south-east Queensland: control of *Tegolophus australis* and *Phyllocoptruta oleivora* (Acarina: Eriophyidae), effect of pesticides, alternative host plants and augmentative release. *Experimental and Applied Acarology* **12**: 195–217.

Index

www.ingramcontent.com/pod-product-compliance
Lightning Source LLC
Chambersburg PA
CBHW052141170526
45159CB00017B/3135